SASQUATCH DOWN

Michael Newton

Typeset by Jessica Taylor,
Cover and Layout by SPiderKaT for CFZ Communications
Using Microsoft Word 2000, Microsoft Publisher 2000, Adobe Photoshop CS.

First published in Great Britain by CFZ Press

CFZ Press
Myrtle Cottage
Woolsery
Bideford
North Devon
EX39 5QR

ISBN: 978-1-909488-37-3

Dedication: To Dmitri Bayanov

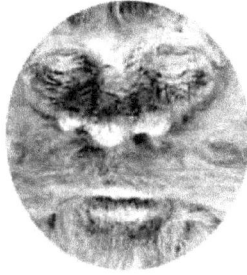

Acknowledgements

T hanks to Jon Downes and everyone else at CFZ Press for their enthusiasm and hard work in bringing this project to life. Thanks also to Dr. Tuklo Nashoba for kind permission to reprint his story of a Choctaw battle with Sasquatch (Chapter 1); to Susie McIntyre at the Great Falls Library and Kathryn Showen and the *Great Falls Tribune* for timely assistance; to *Wenatchee World* librarian Linda Barta, ditto; to Darrell Ehrlic, editor of Montana's *Billings Gazette,* for providing long-lost articles; to Phil Thackeray for sharing his personal story (Chapter 6); to Doug Tarrant for permission to quote from various postings online; and to Dennis Bauer for clarifying sources from his exhaustive Internet archive of Sasquatch shootings (sadly removed from the Web on July 6, 2014, with an announcement that Bauer was "returning to the woods, forests, fields and meadows"). Their assistance is much appreciated, as is my wife's continuing forbearance.

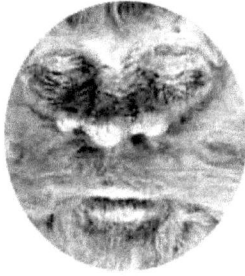

Contents

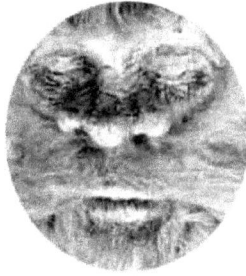

Introduction

Where are the bones?

That question follows Sasquatch researchers wherever they break cover, posed by skeptics and professional debunkers. If large bipedal apes have been reported from six continents throughout recorded history—and they have—why has no one killed or captured a single specimen for scientific study? In our automotive age, why has none been run down anywhere? Why has no hiker, hunter, or explorer found a Bigfoot's bones?

The question, on its face, seems perfectly reasonable. No less a personage than world-renowned paleoanthropologist Louis Leakey, when asked about Sasquatch, retorted, "Show me the bones."[1] And indeed—under the International Code of Zoological Nomenclature and the International Code of Nomenclature for algae, fungi, and plants—each taxon (a group of one or more populations of a distinct organism) is based on a particular "type specimen." In order for Bigfoot-Sasquatch-Yeti to be scientifically classified, named, and (perhaps) legally protected, it must first be proven to exist.

That basic requirement has spawned a fierce "kill/no kill" debate among researchers and monster hunters worldwide. The practical (some say "cold-hearted") pro-kill advocates insist that one specimen must be sacrificed for the good of all, and for the advancement of science. No-kill proponents suggest various alternative methods of proof ranging from photographs or videos—rejected sight-unseen by most skeptics—to plans for collecting flesh, blood or hair for DNA sequencing without harm to a living creature.

So far, neither side has succeeded. Or, have they?

When debunkers ask for bones, they take for granted that none have ever been found, that no cryptic manbeast has ever been killed or captured. The public record, as we shall discover in these pages, flatly contradicts that assumption.

In May 2011, Internet blogger Robert Lindsay wrote that "humans have shot and killed Bigfoots 35 times in the last 182 years"—an average of one kill every sixty-two months since

1829. He also cites multiple cases of manimals struck and killed by trains or cars since the 1880s; lists fourteen alleged discoveries of "Bigfoot burials or possible Bigfoot graveyards" witnessed by humans since 1949; and adds thirty instances of "possible Bigfoot bones, skulls, skeletons or teeth" found since 1858, nearly as frequent as alleged Sasquatch shootings.[2]

Before doubters howl in derision, I am obliged to mention that Lindsay's tally, however wildly speculative or sparsely documented, is in fact *too conservative*. In the following pages, we shall review 165 claims of Bigfoot shootings, either fatal or serious enough to be potentially lethal. Our tally also includes twenty-three cases of supposed Bigfoot roadkills, sixteen alleged observations of burials or "possible" Bigfoot cemeteries, and 107 reports of giant remains discovered worldwide since the 16th century.

Are those accounts reliable? I leave readers to make that judgment, after pondering the documentary and anecdotal evidence. If scientists possessed a Bigfoot/Yeti corpse, the question would be moot.

Or would it?

As our final chapter shall attest, repeated claims have been advanced that a conspiracy exists to cover up proof positive of large unknown primates sharing our planet with mankind. In the United States alone, that plot allegedly includes the military, the Federal Bureau of Investigation, the Smithsonian Institution, various universities, the news media, state police and wildlife officials. The conspirators, we're told, have confiscated cryptid corpses, silenced witnesses, and falsified—or, at the very least, suppressed—reports of unknown primates on the prowl.

Why hide the most astounding scientific find of this or any other century?

According to the theorists, Bigfoot's existence, once established beyond question, threatens science and religion as we know them, challenging both adversaries in the endless struggle between evolution and creationism. Both sides, we are told, have vested interests in purging evidence of manimals at large. A mini version of that conflict drew global attention in 2003, with the discovery of *Homo floresiensis*—nicknamed "the Hobbit"—in Indonesia. More than a decade on, debate still rages as to whether "Flores man" represents a new species of ape, a parallel human race, or simply a group of unfortunates malformed by isolation and inbreeding. Imagine the row—and potential panic—if it were proved that giant apemen have coexisted with *Homo sapiens* from antiquity, their presence concealed by leaders of government and scientific institutions.

Sasquatch Down takes no side in that argument and offers no endorsement of the stories told herein. The text exposes hoaxes where they have been proven, evaluates other reports on the evidence available (if any), and leaves readers to decide the question for themselves. Whatever the final verdict, I hope you enjoy the journey as much as I did.

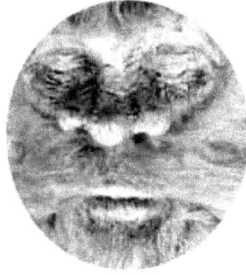

Chapter 1.
Wild Frontier (1767-1899)

L egends and traditions of hairy forest-dwelling manbeasts, known among aboriginal tribes by at least seventy-six different names, were well established when the first European explorers set foot in North America.[1] Our first known report of a white man killing a Sasquatch involves famous frontiersman Daniel Boone (1734-1820), known in equal parts for his real-life derring-do and his addiction to "tall tales." The slaying of a hairy, ten-foot-tall creature that Boone dubbed a "Yahoo" allegedly occurred somewhere in Kentucky, probably during Boone's two-year hunting expedition of 1769-71.[2]

Author Hugh Trotti, writing for *Skeptical Inquirer* in autumn 1994, posits that Boone derived his story of the "Yahoo" from reading *Gulliver's Travels* by Jonathan Swift, published eight years before Boone's birth. Part Four of that popular novel, titled "A Voyage to the Country of the Houyhnhnms," features descriptions of the hairy bipeds and includes Gulliver's capture of a three-year-old male specimen with a "smell very rank"—often featured in modern descriptions of Sasquatch encounters. Boone, in Trotti's view, simply amended Swift's tale and transported the Yahoo to Kentucky for the amusement of his campfire companions.[3]

Against that claim, we have the word of folklorist Leonard Roberts, who reported in 1957 that he had collected "four or five versions" of "a curious and strange legend" from Kentucky's mountains, referring to a Sasquatch-type creature called "Yeahoh," after the sound of its loud, mournful cries.[4] In any case, no extant evidence supports Boone's killing of a Yahoo/Sasquatch in the Bluegrass State.

Our next report describes events allegedly occurring in Georgia's Okefenokee Swamp in June 1829. First reported in Milledgeville's *Statesman & Patriot* newspaper, relayed to us by the *Augusta Chronicle* in March 2000, the tale involves a ferocious "Man Mountain" said to inhabit an enchanted island in the midst of the Okefenokee. Local Indians shunned the island after bloody clashes with "mortals of superhuman dimensions and incomparable ferocity," but whites were intrigued by discovery of humanoid footprints eighteen inches long and nine inches wide, linked to "fearful screams that could only have been made by dreadful monsters."[5]

On June 6, 1829, nine hunters from Georgia and Florida set out to slay the beast, armed with rifles, pistols and swords. Two weeks later, they found giant tracks and pitched camp for the night, prepared to follow the trail at first light. Overnight, however, they were attacked by a "horrible monster covered with hair," seemingly impervious to gunfire. According to the *Statesman & Patriot,* "The huge being, nothing daunted, bounded upon his victims, and in the same instant received the contents of seven rifles. He did not fall until he had glutted his wrath with the death of five of them, which he effected by wringing off the head from the body. Writhing and exhausted, at length he lay upon the ground, with his hapless prey beneath his grasp." When the monster finally died, "wallowing and roaring," the survivors measured its corpse, reporting that it was thirteen feet long "and its breadth and volume of just proportions."[6]

It is a thrilling tale, worthy perhaps of filming as a feature for the SyFy channel, but devoid of specifics including the hunters' names and specific location of the event. No portion of the creature was transported or preserved for posterity.

Blogger Robert Lindsay dates our next story from 1855, though the tale as originally published offers no dates. It allegedly occurred in the southeastern quadrant of Indian Territory (now Oklahoma), created in 1834 and admitted to the U.S. as a state in 1907. The story, presented online by Dr. Tuklo Nashoba, is presented here in full, with permission from the author.

> There are many tales of strange happenings in the forests and woodlands of Oklahoma. Many of the folk have stories about haunted woods, strange beasts out in the woods, and "spooky" noises. There are legends, folktales and family histories where screams in the night have been handed down from family member to successive family member.
>
> Hamas Tubbee was an unusually large man, even for a Choctaw Indian. His father, Hanali Tubbee stood two inches over eight feet in height and weighed five hundred forty pounds. Hamas and his six sons stood about a foot shorter than Pahlumi, or "father" Tubbee.
>
> They were large, exceedingly strong, fierce warriors. Hamas and his sons were the point riders for a troop of Choctaw cavalry known as the "Lighthorsemen." Many in the Choctaw Nation thought it humorous that such large men, riding draft horses, referred to themselves as "Lighthorsemen." Tubbee's men experienced something which none would ever forget.
>
> This day's assignment was to flush out some bandits that had been preying upon the local farmers. A thirty-man troop would be going into an area which later in the state of Oklahoma became the McCurtain County Wilderness Area. These bandits had been not only taking large quantities of corn, squash, and beans, but had as well been taking very young children. This thievery had been taking place across the border in Arkansas as well as in Indian Territory. The captain of the troop of Choctaw cavalry was a man named Joshua LeFlore. Captain LeFlore was of mixed blood, part

French, part Choctaw. The men deeply respected him. Joshua LeFlore was impeccably honest and was brave to a fault.

The men had been traveling horseback non-stop since three o'clock in the morning. They began their assignment at the tribal capital in Tuskahoma and when they finally came to the Clover River, they let their horses eat and the men decided to rest and eat as well. Non-stop riding for eight hours, having to lead their horses across Little River, and the hot July sun were taking a toll on the men and their mounts. When some time had passed, Captain Josh gave the order and the men remounted and they began the last leg of their trip. At or around 4:30 in the afternoon, the troop came to the edge of the area which the bandits were supposed to be inhabiting.

Captain Josh signaled with uplifted hand that the troop should come to a halt. Standing in his stirrups, Captain Josh utilized a ship's eyepiece and promptly turned to his men and gave the command for a full armed charge. The distance between the suspected bandits and the troopers was about five hundred yards. The Tubbee men and captain Josh were at the front of the charge and as the thirty men and he neared the thick, pine forest where the bandits were, two things took place at once...The stench of death assaulted both men and horses, and the horses became uncontrollable.

Horses were rearing, pitching and throwing riders. Captain Josh and the seven Tubbee men were the only ones in the troop whose mounts were disciplined enough that they continued to obey their riders and continued to charge in the midst of the bandits.

When the eight men met with the "bandits" they were totally unprepared for what greeted them. The clearing behind the initial tree cover was actually a large, earthen mound. Strewn about the mound were numerous corpses of human children in varying stages of decay. Most of the bandits had fled, but three really large, hairy ape-like creatures remained at the mound. Captain Josh drew his sabre and with pistol in hand, sabre in the other, charged the huge monsters.

The nearest monster killed Captain LeFlore's horse with one blow of its massive hand. The monster never flinched as Captain LeFlore poured bullets from his Patterson's Colt revolver into the beast's chest. After emptying the revolver into the monster, Captain Joshua continued to press the attack with his sabre. Many times did the sabre meet with the brute's flesh and many times did blood spew from the gaping wounds on the beast's body. So quickly did this engagement take place that the Tubbee men had barely enough time to take aim at the three monsters before one of the beasts flanked the Captain and literally tore off Captain LeFlore's head.

There was not time for any sort of delay due to shock. The Tubbee men opened fire upon the three man-beasts. Seven 50-caliber Sharp's buffalo rifles impacted the three simian appearing brutes at the same time. From years of routine and practice, all bullets smashed into the three monster's heads. Six rounds were fired into the heads of the two monsters which were the culprits that killed their beloved Captain. Only the youngest Tubbee, Robert, had the presence of mind to put a bullet into the head of the third monster.

A legend was born that day. Robert Tubbee, 18 years of age, all six feet eleven inches, three hundred seventy-three pounds of him, chased down a wounded man-beast and finished the beast off with only his hunting knife. By the time the other six Tubbee men caught up with Robert and the monster, Robert had already decapitated the beast. Holding the head aloft with both hands, Robert let out a primal scream which made even the Tubbee mounts panic.

The "Lighthorsemen" gathered their mounts and surveyed what was before them. Absolute carnage littered about the clearing. The partially consumed bodies of nineteen children lay upon and about the mound. The stench of decaying bodies was bad enough, but the over-powering odor of the man-beasts' urine and feces was more than the strongest stomach could endure.

After retching violently, the men of the troop buried the bodies of the children in nineteen small graves, buried their beloved Captain, and as a matter of respect, gave him a twenty-one-gun salute.

They built a large bon-fire, placed the murderous man-beasts upon it, and lit it. As they rode back into Tuskahoma each man struggled with emotions and thoughts he never before imagined.[7]

While granting permission to publish the tale, Dr. Nashoba stressed that it "is just a story" told by his Chickasaw friends, "and probably not fact."[8] No evidence supports it.

British authors Janet and Colin Bord report our next case, published in the *Inyo* (California) *Register* on March 19, 1981, recapping articles from the *Bishop Creek Times*. According to that story, hunters John Clarke, Jack Ferral and Paul Myrtengreen sought to bag a "large shaggy beast" roaming around Round Valley, California. Clarke found it first and tried to lasso the creature while it slept, but the Sasquatch fled, howling and causing Myrtengreen to swoon from fear. On March 25, 1882, Ferral caught the beast feeding and shot it five times at medium range, but it leaped up, roaring, and his horse bolted, breaking two legs as it fled, dumping Ferral in a battered heap. The Sasquatch, presumably, escaped.[9]

In autumn 1886, a truly monstrous Sasquatch was killed in Maine, according to the now-defunct *Waterville Sentinel*. The story begins with "an affrighted Frenchman from over the line"—i.e., in Québec—who appeared in Elm City weaving a tall of terror. The *Sentinel* reported: "The Frenchman's story, which is implicitly believed, is that three men were camping out in the woods about a hundred miles north of Moosehead Lake. Two of the campers were away from the camp for a week and came back to find the dead body of their companion." Frightened and furious, the survivors "went for help and reinforced by a dozen others searched the woods for the unknown murderer. It proved to be a terrible wild man, ten feet tall, with arms seven feet in length, covered with long, brown hair. The party fired several shots at him and finally succeeded in reaching a fatal spot, laying the monster low." The Frenchman did not claim to be a witness, but he still "had his fellows in town all by the ears."[10]

Strange as it was, the story gained wide currency, repeated by the *Wilton Record* on October 6, 1886; by Bangor's *Industrial Journal* on October 8; and by Vermont's *St. Albans Daily Messenger* on October 12. Papers who ignored the tale include the *Bangor Whig and Courier*, the *Bangor Commercial*, the *Piscataquis Observer*, the *Aroostook Times*, and the *Houlton Pioneer Times*.[11]

Internet blogger Dennis Bauer reports our last case from the 19th century, claiming that sometime in the "late 1800s" a moonshiner in Winston County, Alabama, shot a Sasquatch that was following his mule-drawn whiskey wagon.[12] No further details are provided, and Bauer's hyperlink to an alleged supporting source was dead as this book went to press.

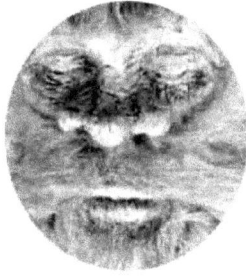

Chapter 2.
Shoot First, Ask Questions Later (1900-1945)

Reported Sasquatch killings escalated in the 20th century, with eleven incidents reported between 1900 and the end of World War II. Our first account comes from Prince of Wales Island, largest of 1,100 islands in the 300-mile-long Alexander Archipelago of Southeast Alaska. Slightly larger than the state of Delaware, at 2,577 square miles, Prince of Wales is mountainous, most of its land covered by the Tongass National Forest. Somewhere in that wilderness, unidentified Inuit tribesmen allegedly killed a "hairy man" sometime in 1900, then buried its corpse on a beach, location unknown. Researcher Raymond Crowe, while logging the report twice in his International Bigfoot Society (IBS) database, rightly rates the event as "inconclusive."[1]

Five years later, another Sasquatch was reportedly killed at Gardner Canal, British Columbia. Gardner Canal is, in fact, a side-inlet of the larger Douglas Channel, fifty-six miles long. The British Columbia Scientific Cryptozoology Club (BCSCC) notes "reports" of a Sasquatch killing in the neighborhood, in 1905, but admits that "there are no details in regard to this event."[2]

Our next report also comes from British Columbia, badly garbled in translation on the Internet. The BCSCC's website, citing an otherwise undescribed 1963 article from *British Columbia Digest,* claims that "Two generations ago, a Kitimat native shot a Sasquatch. Several other Sasquatches appeared and gave the impression that they were about to attack the man as he attempted to drag the body out. He ultimately fled to his canoe and it says much about the powers of restraint of the BC Sasquatch that none of them attempted to pursue the man any further."[3] Blogger Robert Lindsay reports the same shooting in abbreviated form, citing the BCSCC posting, but oddly dates the incident from 1965.[4]

Further research proves both accounts are erroneous. The full story, reported by Bill Oliver in December 1998, reads as follows [spelling and punctuation corrected]:

I asked if there had been any Sasquatch reports in the area and he said he knew of a story of a man who had shot one. He passed on the number of someone who knew the story better. I called and spoke to an elderly man, Ken, who first asked me why I wanted to know. I was lucky, for it was this gentleman's grandfather who was reported to have done the shooting. The first thing Ken told me was it irritated him how the story had changed over the years since it happened. In particular, it disturbed him that people said his grandfather shot a Boqwish and that maybe this was an opportunity to tell it like it happened.

In the spring of 1918, William Hall was out hunting for the family's needs with his good friend. In this case he was bear hunting eight miles west of Kemano in a small area known as Miskook, a small inlet on the Kemano River. His friend and himself were joined by an elder whose job was to wait in the canoe and watch the supplies. As William and his friend made their way through the terrain they came upon a split in the valley. It was here where they separated their ways. As the custom went, a wooden stake was pounded into the ground. Upon return, the first hunter would remove it and lay it on the path to let the other know he had safely arrived and to meet him down at the river's edge. William, being the first back, did so and started his way back to the waiting elder. It was here on a small trail he came upon a group of four Sasquatch, or as known to the Kitimat Indians, the Boqwish. In absolute terror he started to run, but apparently blacked out. When he came to, he found himself on a large rock. The four Boqwish were below, reaching out and attempting to grab the startled hunter. In his own native tongue he spoke to them and said that he was not there to harm them but only hunting for food for his family. It was at this time that the aggressors seemed to back off, as if they understood. He made his way off the rock and began back to the river's edge where his partner had been waiting in the canoe. Along the way the creatures continued to follow him to the river and now the waiting elder also said, in his language, that they were not out to harm them. Again, seeming to understand, they left. Upon getting into the canoe William Hall slipped into a coma that lasted four days. It was on the fifth day he awoke. It is well reported that accompanying the bigfoot is a foul odor that fills the air whenever the creature is near. William had the same rancid odor permeating his body until the day of his death, eight years later. So bad was it that he built a hut for himself to sleep in, so as not to offend his family. Since the day he came out of the coma, Ken said, "My grandfather could foresee the future." He displayed other traits of a psychic nature as well.[5]

In short, no shooting, whether the event occurred or not.

Flash forward to 1921, in Terrebonne Parish, Louisiana. Robert Lindsay reported this case online, crediting it to an unspecified issue of Ray Crowe's *Track Record* newsletter, while managing to garble both the journal's title (rendered as *Bigfoot Track Record*) and the parish's name (spelled "Terrebone"). According to Lindsay, "Hunters killed a Bigfoot and dumped the body in an old well. Later a skeleton was found in the well and taken to the Tulane University anthropology department, where it disappeared. The anthropologists were not able to identify the skeleton and were mystified by it." He goes on to say, "This is probably one of the best verified cases of the killing of a Bigfoot. After the skeleton was found in the well, many of the local college boys from Terrebone [*sic*] and Tulane University came around and took photos of it. Residents of Terrebone said that for many years afterward, as the college boys grew up, many of them still had photos of the Bigfoot skeleton."[6]

Alas, nine decades later and counting, those photos—if they existed—have gone the way of the vanishing skeleton, leaving us with nothing.

Ape Canyon
No Bigfoot tale of the early 20th century is better known than the supposed battle of Ape Canyon, on the Plains of Abraham near Washington's Mount St. Helens, in July 1924. Widely described in countless books and articles over the years, the incident was memorialized thirty-three years after the fact in a booklet dictated by key participant Fred Beck, self-published in September 1967.

Beck opened by telling his readers, "This is not a large book, but may the largeness be conveyed by the picture I hope to paint of truth."[7] Beck and four friends were working their Vander White goldmine for the sixth consecutive year, when they began hearing shrill whistles and a "booming, thumping sound" from the forest at night, as of some great ape beating its chest. On the night in question, while fetching water from a nearby spring, Beck and another miner fired shots at a seven-foot bipedal creature covered with blackish-brown hair. Both missed, but the beast found reinforcements and attacked their cabin after nightfall, hurling boulders and tree branches at the windowless abode. The miners fought back until daylight, firing through gaps in the logs, then cautiously ventured outside, where Beck saw one of the monsters standing on a nearby cliff. His next shot sent it plummeting into a deep ravine.

From that sketchy account, Beck goes on to say, "The purpose of this book is not only to relate my experiences, but also to bring to light my knowledge about the Abominable Snowmen. I do not wish to embark upon an expedition, but I wish to tell what these beings are." And according to Beck, what they are is a tribe of "psychic" creatures "not entirely of the world." Events leading up to the battle were "filled with the psychic element," beginning with spiritual visions during Beck's childhood and "many years in healing work" for the Seventh-Day Adventist church. "The method we found our mine was psychic," Beck claimed, led by "a spiritual being, a large Indian dressed in buckskin," whom Beck and his partners called "the Great Spirit." The apemen, Beck said, were "lower or grosser manifestations" of the same supernatural energy. When huge footprints first appeared on a nearby sandbank, with a stride "160 feet long," partner Hank declared, "No human being could have made these tracks, and there's only one way they could be made. Something dropped from the sky and went back up."[8]

In conclusion, Beck opined, "The Abominable Snowmen are from a lower plane. When the condition and vibration is at a certain frequency, they can easily, for a time, appear in a very solid body. They are not animal spirits, but also lack the intelligence of a human consciousness When reading of evolution we have read many times conjecture about the missing link between man and the Anthropoid Ape. The Snowmen are a missing link in consciousness, neither animal nor human. They are very close to our dimension, and yet are a part of one lower. Could they be the missing link man has been so long searching for?"[9]

He closed with a pessimistic prediction for Sasquatch hunters: "I hope this book does not discourage too much those interested souls who are looking and trying to solve the mystery of the abominable snowmen. If someone captured one, I would have to swallow most of the content of this book, for I am about to make a bold statement: No one will ever capture one, and no one will ever kill one—in other words, present to the world a living one in a cage, or find a dead body of one to be examined by science. I know there are stories that some have been captured but got away. So will they always get away."[10]

Beck's memoir makes it tempting to dismiss the whole Ape Canyon story as the ravings of a religious zealot, but it appears that something actually *did* occur at the Vander White mine. Author Mark Sumerlin offered this account in the August 2002 issue of *Fate* magazine.

> During the 1970s, books and movies turned attention to Bigfoot as never before. In 1979, on the heels of all of this, Rant Mullens, a Washington State woodsman who was born in 1896, stepped forward to tell how he had done much to promote the Bigfoot legend by his prankish hoaxing decades earlier. In fact, according to Mullens, he had all but created the Bigfoot character single-handedly ... literally, the way he described it.
>
> It all started in 1924, before the names "Bigfoot" and "Sasquatch" had caught on, and the legendary primates were seemingly known as just "apes." Mullens and a friend were hiking back from a fishing trip in Washington's Spirit Lake region when they paused to prankishly roll some stones onto a miners' cabin in the Ape Canyon region. The miners "came the following day to the Spirit Lake Ranger Station and said that hairy apes threw rocks at them," Mullens recalled with relish, adding, "they had all the lawmen up there looking for the apes."[11]

A very different tale from Beck's—but is it true? Mullens, an admitted inveterate hoaxer, later carved crude wooden feet in various sizes and boasted of planting fake Sasquatch tracks in Washington State, also claiming he sent a pair to notorious California track-faker Raymond Wallace. For a detailed discussion of Wallace, see my previous book *Hoaxed!* (CFZ Press, 2013). For now, let it suffice to say that no aspect of the Ape Canyon case—including the role of Rant Mullens—is clearly established.

Between World Wars

Our next reported incident, from 1928, is typically vague. John Green were first to publish the account, described as follows fifty years after the fact:

> George Talleo, from Bella Coola [British Columbia], had shot at a Sasquatch on a hillside above South Bentinck Arm back in 1928. He said he was on his trap line when he saw the thing suddenly stand up from behind the trunk of a fallen tree. He snapped off a shot at it with a small-caliber rifle, and it somersaulted over backwards, out of sight. George also decided he had urgent business elsewhere, and he never went back until 1962.[12]

Readers may recall that I promised to omit cases of gunmen simply "shooting at" Sasquatch, with no proof or claim of a kill. I include Talleo's sighting here because blogger Robert Lindsay has converted it to a slaying, with the wholly unsupported claim that Talleo "shot and killed a Bigfoot."[13]

Nelchina, Alaska, is allegedly the locus of our next Sasquatch slaying, vaguely dated from sometime in the 1930s. The Sasquatch Tracker blog online, self-billed as "Alaska's Cryptid Authority Since 2005," says simply: "Native man allegedly kills Sasquatch, no other details."[14] The blog credits its terse report to Bobbie Short's Bigfoot Encounters website, where we find the tale reversed, claiming that Sasquatch—or, rather, a "cannibal giant" called Gilyuk—actually killed an Indian.[15] That account, in turn, originates with author Russell Annabel, writing for *Sports Afield* magazine in 1956.[16]

Robert Lindsay is our sole source for the next alleged Sasquatch slaying, though he credits deceased Bigfoot researcher Datus Perry for the original tale. The event allegedly occurred in 1937, along the Green River in Washington State. Lindsay writes: "In the Cascades east of Tacoma, a hunter saw a bear grubbing in a log and shot and killed it. Turned out he had killed a Bigfoot. Feeling that he had shot a 'hairy man' (a human being), he buried it under a pile of rocks and never told anyone until he confessed on his deathbed."[17] Lacking a name or any further details, the incident remains unverified.

Our next incident—or is it two?—ranks among the most confused from the dead-Sasquatch roster. Field researcher Dennis Bauer leads off, writing: "1940-00-00; MO, unknown; bigfoot killed after it ripped apart a cow."[18] Further digging brings us back to Bigfoot Encounters online, where we read:

Missouri Mo Mo tale: No date or location
Jared Sparks in the southeastern swamps of Missouri shot and killed something he could not identify that was described "something like a gorilla" According to Sparks, the creature was able to rip up full grown cattle and horses...

Southeast Missouri: 1940
An animal was reportedly shot to death after ripping a cow apart in the Nigger Wool Swamps. May be a follow up article to the Jared Sparks story above.[19]

Bobbie Short's source for the latter story is author John Keel, who actually credits Jared Sparks with the shooting at Nigger Wool Swamp (not "Swamps"), near Morehouse in New Madrid County, no longer found in modern gazetteers. Keel, however, dates that shooting from "the late 1940s."[20] Robert Lindsay reports the Sparks shooting as a single case, but leaves the date as 1940.[21] Janet and Colin Bord report Keel's basic information, without naming Sparks, then suggest that the incident "may be the same case as Piney Ridge 1947"—referring to a fruitless hunt for a goat-killing Sasquatch in Pulaski County.[22] On balance, it seems that there was only one shooting, if in fact there were any, which left us no evidence.

"Holy Buckets!"
We have more details for our next case—but again, no proof. The incident occurred, we're told, in November 1941, when a seventeen-year-old poacher known only as "Peter" was hunting moose without a license near Basket Lake, approximately fifteen miles from Gypsumville in north-central Manitoba, Canada. Although kept secret for decades, the story was finally aired in 2004, thanks to researcher Curt Nelson from the Bigfoot Field Researchers Organization (BFRO). Nelson interviewed Peter in May 2003 and published his story on the BFRO's website in October 2004.[23]

Nelson's report begins with some minor confusion, first dating the event from November 16, later writing that it occurred "in the first week of November." After hours of stalking, Peter thought he'd found a moose: "I didn't see a calf, and no horns, so I knew it was a cow." He fired one rifle shot, but did not kill the creature outright, being forced to follow its blood trail for half an hour through thick brush. Finally glimpsing his prey again, at a distance of forty-five yards, he fired a second time and brought it down. Approaching cautiously, in case the wounded animal was still alive—

> I looked, I could see him...what the hell is this? Holy buckets! He's lying there and one foot was up, you know...So I nudged him in the foot and slowly walked on this side, still hanging onto my rifle like I was supposed to, and I picked the hand up with my right foot, to see the bottom. And I walked around and I could see where I hit him, in the back, high in the back, between the shoulder blades, right in the back. It must have been bent over...because—to look at the moose track and the blood or something like that—I didn't see a head.

From that, it sounds as if Peter believed his first shot hit a moose, in fact. If so, he never saw that animal again. Walking around the corpse, he thought, "Holy God, what the heck am I gonna do now?" Suddenly frightened, by the creature and the consequences for himself—a teenage poacher and a child of German immigrants in Canada, at war with Germany—Peter departed from the woods with all dispatch, leaving his kill behind.

Interviewed at age seventy-nine, Peter guessed that the Sasquatch was eighteen to twenty-four inches taller than his own six feet, with "very large" hands bearing four fingers and an opposable thumb. The creature's five-toed feet were "very flat," fifteen to sixteen inches long, with bare soles, "but they weren't white...He's not a very hygienic creature, you know, he doesn't wash or anything, so his skin is dirty and a little brown, but white like my dirty hands." Its dark brown pelage looked "like hair, not like fur," around eight inches long on head and shoulders, shorter over the rest of its body. Peter's initial fear was that he had gunned down "a hermit type person."

A quarter century later, viewing the famous Patterson-Gimlin film from California, Peter says he finally admitted to himself that he had killed a Sasquatch. He began to speak about the incident with friends, relieved to hear that some of them had sightings of their own to share, but when talking to Nelson he retained "the attitude that since he can't prove it happened, his story isn't really worth telling, that it's just another unsupportable claim." On a second visit, when he led Nelson to the approximate site of the shooting, sixty-two years and local logging operations made the spot impossible to find.

A World at War

America's belated entry into World War II eclipsed most local news in the United States, but we have two cases of reported Sasquatch killings from the war years. Curiously, both date from 1943 and come from neighboring southern states.

Dennis Bauer provides the first case, reading in its entirety: "1943 winter; AL, Clarke; human shoots a bigfoot."[24] Alabama's Clarke County covers 1,253 square miles in the southwestern part of the state, but Bauer offers no specific pointers. "Winter," of course, may refer to January, February, early March, or late December. No other published source refers to this killing, and while Bauer provides a link to his Internet source, that link is dead today.

We have more details for the next supposed incident, though Bauer's is typically terse: "1943-00-00; GA, unknown; bigfoot kills livestock, humans killed the bigfoot."[25] Robert Lindsay does better, placing the event near Georgia's border with South Carolina, citing his source as Ray Crowe's *Track Record* journal.[26] Crowe, in turn, pegs the source of the story as author John Green, who writes:

> The Georgia story was told to Rich Grumley, of the California Bigfoot Organization, by a man who claimed to have seen the creature after it was dead, when he was a boy in 1943. He said the creature was killing sheep and calves by tearing off a leg and leaving the animal to die. Men tracked it onto a small mountain, cornered it and felled it with some 60 slugs from their shotguns, one of which went through an eye into the brain. Brought to town in a pickup truck, it was so big that its torso was wider than the truck box, and even though the tailgate was down its feet dragged on the ground while its head was propped up against the back of the cab. It was covered with reddish-brown hair on the head and chest. The smell was terrible. The man said that the creature was buried under a pile of rocks, and Rich, who knows the town where it is supposed to be, would like very much to go there and look under the rocks.[27]

Green does not name the town or its county, nor does he seem to know if Grumley (deceased since 2000) ever made the trip from California, but Green personally dismisses the tale on grounds that a Sasquatch killing should have been front-page news. We must take Green's word that it was not, since a dearth of geographical pointers makes further research fruitless.

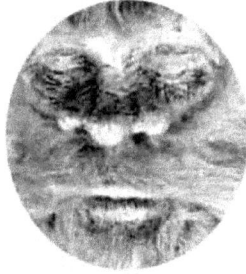

Chapter 3.

Cold War, Hot Lead (1946-1969)

The quarter century following cessation of hostilities in World War II produced twenty-seven reports of Sasquatches killed or seriously wounded in North America. The first account, from Dennis Bauer, reads: "1940's late; OR, Coos; human shoots a bigfoot seen chewing on live cow."[1] This time, Bauer's hyperlink to his source functions perfectly, leading us to the BFRO's website. According to that story, filed in May 2004, a betrothed couple saw Sasquatch near the Coquille River in Coos County, Oregon, and later told the young man's father about their encounter. That, in turn, evoked a flashback to August 1949.

> His father told us of the rancher at the foot of the Capes (also on Hwy 101) who had been riding to check on his cattle when he heard a cow bellowing in agony. His horse became nervous but he forced it on and found a very large hairy animal chewing on the live cow. He carried a 30.06 rifle and shot the creature. It stood up and ran off on two legs. He followed until he lost the trail of blood in the rocky terrain.[2]

No body was recovered and, of course, DNA testing was unknown at the time.

A second Oregon story from 1949, otherwise undated, also involves a Sasquatch attacking livestock. Ray Crowe's IBS database reports that the incident occurred on a farm near Oregon City, in Clackamas County, where "a Wild Man was caught eating raw albino turkey. A neighbor shot the Wild Man and drew blood, but it jumped over a fence and ran away."[3] Again, no kill. No evidence preserved.

Dennis Bauer presents our next report, reading: "1950-00-00; PA, Indiana; humans shoot a bigfoot."[4] There *is* a town called Indiana in the state of Pennsylvania, the seat of Indiana County, but Bauer's listed source—Rick Berry's *Bigfoot on the East Coast*—does not contain the information found on Bauer's website. Berry's only report of a Pennsylvania shooting for this decade reads as follows:

> 1950's Cambridge Springs. A group of people waited on a porch one night for a creature that had been banging on an old farmhouse. A little past midnight, they saw an 8 to 10 foot tall, bipedal creature covered with thick fur. The men shot at it and it ran through a cornfield and into the woods.[5]

Robert Lindsay offers our next alleged Sasquatch slaying. He writes online: "1953: Alder Creek Canyon, Sandy, Oregon. East of Portland, a hunter shot and killed a Bigfoot, then buried the body. Reported by Peter Byrne."[6] Ray Crowe echoed that claim, sans mention of Byrne, writing: "There was a 1953 report of a hunter shooting, killing, and burying a Bigfoot in the Alder Creek Canyon."[7]

It is entirely possible that Peter Byrne, a lifelong Yeti-Sasquatch hunter, did in fact allude to such an incident as some point in his fifty-year career of tracking cryptids, but it does not appear in his 1975 book on Sasquatch, titled *The Search for Big Foot*. Furthermore, Janet and Colin Bord have put a nonviolent twist on the case. Their summary of the event (unedited), reads:

> Summer 1953/Alder Creek canyon, nr Portland, OR/Middle-aged man/While fishing apart from 2 companions, saw huge, hairy Bigfoot watching him from thicket/ Unidentified newspaper report of 5 Aug. 1963, probably a Portland, OR, paper, repr. BSIS vol. 2 no. 4 p. 5.[8]

"BSIS" refers to a newsletter published by the long-defunct Bigfoot/Sasquatch Information Service, featuring photocopied articles from various newspapers, frequently (one might say inexcusably) unidentified.

Three years later, Dennis Bauer claims, a California hunter "possibly" shot a Sasquatch at or near Shasta.[9] He cites the IBS database as his source, but no corresponding report appears in the collection presently viewable online.[10] Leaving aside the "possibly," Bauer's skimping on geographical details foils any further research. There is a town called Shasta in California's Siskiyou County, together with a mountain of the same name and a nearby city called Mount Shasta. Attempts to trace this ephemeral report proved fruitless.

Bauer reports another ambiguous case, this time from Oregon, occurring in 1957. His brief account reads: "1957-00-00; OR, Deschutes; bf steals shot deer, hunter shoots bf, bf escapes with deer."[11] Again, he cites the IBS database as his source, but the thirty-six reports presently listed from Deschutes County include no shootings or theft of dead deer.[12] However, author John Steele presents a strikingly similar case from Washington State, occurring the same year.

> In 1957, at Wanoga Butte, Washington, Gary Joanis and Jim Newall were hunting, and had just shot a deer. Before they could get to it, a 9 foot tall, hairy creature walked into the clearing, and picked the deer up, and carried it off under its arm. Joanis, annoyed about losing his deer, fired several shots with his 30.06 into the

creature's back, but the creature never stopped walking. However, it did emit a "strange whistling scream." Perhaps a cry of pain?[13]

Dennis Bauer logs our next report: "1958 or 1960; TN, Overton; farmer kills bigfoot."[14] His hyperlink takes us to the Gulf Coast Bigfoot Research Organization's (GCBRO) website, where we find a long report from Tennessee researcher Mary Green. Green's anonymous witness recalls an incident from childhood, "around 1958-60," occurring near Livingston in Overton County. There, we find that Bauer has misspoken. The original (unedited) report tells us:

> I was real young and can't remember a lot of details (but my sister will be here Thanksgiving and will tell me the whole story. She is a Christian and would not lie about the Bear Man). My grandfather, who often walked all over his mountain acres discovered the dead remains of a unknown creature. He didn't know what quite to make of it. He called my father (his son) who was a teacher at the local high school, to come quick. I remember all of us kids running to the car and driving up there with dad. [Grandmother] wouldn't let us get near it and I still don't know what ever happened to the remains. I think I did get up near it. In the back of a 4 or 5 year old's mind...I still recollect looking at a Bear Man with no clothes and lots of hair. I remember the smell was so bad, you could smell it all the way to the farm house. The creature was found in a field up near a tree line about ¼ mile from the grandparents' house.[15]

As written, this case properly belongs in Chapter 8, with other alleged discoveries of Sasquatch remains, but I include it here since Bauer chose to log it as a killing.

One more case completes our roster for the 1950s. Dennis Bauer's online list tells us: "1959-00-00; OR, Douglas; boy shoots bigfoot seven times."[16] His hyperlink leads us to Oregonbigfoot.com, a priceless website established by researcher Autumn Williams, whose files are by no means restricted to Oregon. There, we read:

Date: October 1959
Douglas County, OR
Nearest town: Tenmile
Description of event: Black BF came up hill after two boys, chased them along ridge. They said at least 12' tall. Apelike face, very heavy, bowlegged. Seemed to be herding them away rather than trying to catch them. Arms outspread. Very long hair on forearms. Older boy shot it seven times with a 30.06. It slumped twice until its knuckles hit the ground but kept coming and didn't seem to lose its temper. It screamed, "like a cat, but louder." Younger boy also shot with a smaller gun....The boy who did the shooting was considered an expert hunter.[17]

Perhaps, but neither of the young gunmen—named by Janet and Colin Bord as Wayne Johnson and "Waiter Stork"[18]—managed to bring the creature down, despite seven shots from a powerful .30-06 rifle—America's standard-issue military rifle cartridge from 1906 to 1962—plus more from an unidentified smaller-caliber weapon.

Robert Lindsay opens the "swinging" 1960s with an undated report that reads: "Douglas, Oregon: In the Cascades west of the Umpqua National Forest, a farmer shot a Bigfoot and then somehow managed to take it back to his house, where he left it outside. Other Bigfoots then came that night and retrieved the body."[19] He credits Ray Crowe's *Track Record* as a source, also undated, but no such case appears in the IBS database for Oregon sightings.[20]

The Iceman Cometh

The "Minnesota Iceman" case is arguably the most famous case of a purported Sasquatch slaying from the 20th century. Most readers will be familiar with basic details of the case, but for any new arrival on the scene, a brief summary is in order.

During 1968-69, showman Frank Hansen toured the North American carnival circuit, hauling a freezer said to contain "The Mysterious Siberskoye Creature," advertised as something "prehistoric." No one paid much attention until December 1968, when herpetologist Terry Cullen viewed the exhibit at Chicago's International Livestock Exposition and phoned Dr. Ivan Sanderson at the New Jersey headquarters of his Society for the Investigation of the Unexplained, suggesting that the relic might be genuine. According to Cullen, Hansen claimed the body was found floating in a block of ice, somewhere off the Siberian coast, and passed through several wealthy hands until an unnamed "California millionaire" offered to let Hansen tour with the corpse for two years. The owner shunned publicity, saying, "I don't want to die and go down in history as the man who upset the biblical version of creation."[21]

Sanderson traced Hansen to his Minnesota farm, dropping in for a visit with Dr. Bernard Heuvelmans, the "father of cryptozoology." Over the course of three days they examined the body as best they could, in its tomb of ice, taking measurements and many photographs. In February 1969, Heuvelmans published his "Preliminary Description of the External Morphology of What Appeared to be the Fresh Corpse of a Hitherto Unknown Form of Living Hominid" in the *Bulletin of the Royal Institute of Natural Sciences of Belgium* (Vol. 45, No. 4, February 1969). Sanderson, while working on a scientific paper of his own, initially aired his findings in *Argosy* magazine's May 1969 issue. Dubbing the creature "Bozo," he described it in detail, concluding:

> Let me say, simply, that one look was actually enough to convince us that this was—from our point of view, at least—"the genuine article." This was no phony Chinese trick, or "art" work. If nothing else confirmed this, the appalling stench of rotting flesh exuding from a point in the insulation of the coffin would have been enough... [T]he proportions of this body, and several of its special features, are just not known at all—or, at least, have never been suggested either by paleontologists who have studied the fossil bones of primitive man-things, or even by the skilled artists who have fleshed out and made constructions of what the former have found. In fact, any "artists" setting out to "make" such a thing would have had to have a model, and none is available. But, apart from that, you can't completely fool two trained morphologists with zoological, anatomical and anthropological training. No! Bozo is the genuine article.[22]

Sanderson repeated that endorsement, with more elaborate details, in a scientific paper published on June 18, 1969. Dr. Heuvelmans, meanwhile, was bold enough to proffer a formal scientific name for Hansen's specimen: *Homo pongoides* ("apelike man"). Five years later, the "Minnesota Iceman" served as Exhibit A in Heuvelmans's latest book, *L'homme de Néanderthal est toujours vivant* (*Neanderthal Man is Still Alive*). Sadly, despite repeated predictions of its imminent translation, that work remains unavailable in English.

Meanwhile, Frank Hansen wrote an article for *Saga* magazine—and his story changed radically. Published in July 1970, Hansen's latest version of the Iceman story cast him as the creature's slayer. The thing was not Russian after all, he wrote, but hailed from the vicinity of northern Minnesota's Whiteface Reservoir, where Hansen shot it—in self-defense, of course—in autumn 1960. Fearing prosecution on some unknown charge, Hansen stashed the body in his freezer for nearly a decade, until he felt it safe to put the body on public display.[23]

Life was sweet and profitable until Sanderson and Heuvelmans arrived on Hansen's doorstep and revealed his secret to the world. Fortunately, Hansen had taken the precaution of commissioning a life-size model of his trophy in advance, before he started touring, and in the hectic spring of 1969 he switched the bodies, on advice from his attorney, spreading word that its wealthy owner had demanded the creature's return. After touring a while longer with the fake corpse, billed as "a fabricated illusion," Hansen retired the model and went back to showing antique tractors at fairs.[24]

Bigfoot blogger William Jevning stirred the pot in December 2011, with an online story adding new details to the Iceman saga. Jevning's source was longtime Sasquatch researcher and conspiracy theorist Doug Tarrant, whose (uncorrected) email on the subject read as follows:

> The Smithsonian sent Ivan T. Sanderson and Bernard Heuvelmans to Frank Hansen's farm to inspect and take photographs of the iceman thru the ice. Having been shot.... the FBI wanted to also check it out in case it was a wild man (human) and a possible homicide occurred. It was when the FBI got interested, that Frank panicked and took the cadaver up into Canada and disposed of it. When Frank couldn't get his refrigerated truck back into the USA from Canada, Senator Walter Mondale (of Minnesota) was asked to help of which he did but wouldn't comment further as to what he knew. I have that letter in my files. The plot thickened and ended there. Covered up?? Frank then had a couple latex/rubber models made and the carnival toured with them. When the FBI inspected the fake models, they lost interest and wrote it off. The article in ARGOSY magazine was screwed up with details. The late B. Ann Slate, who ... wrote for ARGOSY and FATE magazines and said that all the details were off base. All wrong. So if I've shed some light on all the confusion, then good![25]

In fact, Tarrant only made the story *more* confusing. To address his points in order:

1. No evidence exists to indicate that the Smithsonian Institution "sent" Sanderson and Heuvelmans to Hansen's farm. All identified parties agree that Terry Cullen first called Sanderson, who then contacted Heuvelmans.
2. The Federal Bureau of Investigation, although well known for meddling in local affairs

and violating the various laws it is tasked to enforce, has no jurisdiction in homicide cases unless the slayings occur on federal property or involve specific victims: presidents, members of Congress, or U.S. government employees. FBI investigation of terrorist killings is a recent addition to the list, and has no application to the Iceman's death in any case.

3. Senator Mondale may indeed have helped a constituent resolve some Canadian border-crossing conundrum, but there is no obvious reason why Hansen could not have crossed the border unobstructed, as countless other refrigerated trucks do on a daily basis. Why would Canadian Customs officials detain an *empty* trailer?

4. Finally, while any magazine article may go to press with "screwed up details," Tarrant offers no specifics or explanation for his statement, and the fact that Ann Slate—coauthor of *Bigfoot* (1976)—occasionally wrote for Argosy would not make her privy to editorial foibles in Sanderson's story from 1969.

There matters rested until June 2013, when Steve Busti—owner of the Museum of the Weird in Austin, Texas—allegedly bought the "original" Iceman from Hansen's estate. Two years of research had convinced Busti that Hansen told the truth in *Saga,* four decades earlier. "He shot it in Wisconsin," Busti told the *Huffington Post*. "Its eyeball's blown out and its arm is broken. I couldn't believe it had been in Minnesota the entire time."[26]

Conclusive proof—if the carcass exists. Thus far, if any scientific study has been undertaken, its results are tightly under wraps.

Fire at Will
Around the same time Frank Hansen allegedly killed the Iceman in Minnesota, during winter of 1960, hunters Samson Duncan and Timothy Robinson claimed a similar encounter at Watson Bay, on British Columbia's Roderick Island. According to Janet and Colin Bord, they "Shot at small Bigfoot on beach and found blood on snow where it had been, but were afraid to follow the creature."[27]

Dennis Bauer's next contribution reads: "1963-00-00; SC, Cherokee; human shoots a bigfoot."[28] He links readers to an article from the *Arkansas Democrat,* posted on the BFRO's website, which includes no mention of South Carolina.[29]

That newspaper article *does* support Bauer's next citation, which reads: "1965-00-00; AR, Miller; human shoots a bigfoot."[30] The shooter in question, we're told, was fourteen-year-old James Crabtree, who met an eight-foot, reddish-haired creature near Fouke, Arkansas, and "shot the creature three times in the face, but with no effect."[31] Three blasts from a *shotgun,* no less, which should have decapitated any man-sized beast of flesh and blood on the spot. The result: "As the creature didn't seem to notice, the boy left as fast as he could."[32]

Dennis Bauer posts two more exciting tidbits from 1965. The first reads: "1965-00-00; WA, Yakima; boy shoots bigfoot, bigfoot tore boy apart and crushed his rib cage."[33] This tale does not strictly meet our criteria, since the Sasquatch escaped after slaying a human, but a slaughtered boy should certainly be newsworthy. Bauer sites the IBS database as his source, and while that database online offers a matching summary—"BF CHASED GIRL, BF SHOT AT, TORE BOY

APART"—the "full report" it promises is found nowhere online today.[34]

Bauer's second story, also credited to the IBS, reads: "1965-00-00; WY, Teton; coyote hunter shoots bigfoot in chest, it dies they abandon it."[35] Ray Crowe, in turn, credits the tale to Peter Byrne, describing the event as follows:

> Seems two hunters from Ohio went to a spot near Jackson Hole, Wyoming, in 1965 and were road-hunting...shooting from the window of their car. They saw a coyote and shot it, but did not kill the animal. Following it over a small embankment, they encountered a Bigfoot and promptly shot it in the chest with a .30-06. The Bigfoot fell over on its back and lay there quivering; its fingers opening and closing and the eyes opening and closing. On closer investigation, they noticed the thing looked human and were sure they had shot some farmer's handicapped child, and fearing a police charge...they got back into their car and headed back to Ohio. It was seven years before they told anybody, and then the story spread. Peter tracked the story back to its source, returned to the area with one of the men and spent a week in 1972 searching for bones and teeth with Patry Hull as an assistant. Nothing was found.[36]

> John Steele presents our next case online. It reads:

> In Washington, during the summer of 1966, there were numerous sightings of a white/gray Bigfoot, 8 feet tall with red-eyes, weighing at least 600 pounds and walking like a human. A group of men often went looking for it, and usually [?!] found it in a gravel pit. Roger True fired at it from a range of only 20 feet, and hit it at least three times from his .270 rifle, but didn't knock it down. Tom Thompson fired his 10-gauge shotgun from 10 yards. He said it "screamed, a sort of high-pitched squeal," but the shots didn't stop it running away.[37]

We can only share Steele's amazement. The Winchester .270-caliber cartridge, introduced in 1925, is a popular hunting round. As described by its manufacturer, "When loaded with a bullet that expands rapidly or fragments in tissue, this cartridge delivers devastating terminal performance."[38] Ten-gauge shotguns are the largest caliber readily available to U.S. "sportsmen," the bore measuring .78 inches (19.7mm) in diameter. It delivers a savage blast at close range, formerly sold to law enforcement as a "roadblocker," with a promise that it would stop fleeing getaway cars.[39] Either weapon should have done for Sasquatch, yet both allegedly failed.

Bobbie Hamilton posted the next alleged Sasquatch slaying on *Bigfoot Encounters*.

> Kobuk River, Alaska, 1966: Jean Joiner of Nome, while at his mine near Dall Creek on Jade Mountain often found large manlike tracks. One day he finally faced the creature and shot an "upright walking bear" in the back, killing it. Joiner found the thing looked so human he didn't know what it was, afraid, he cut it up and threw it in the stream. Bob Larson, Bureau of Land Management area manager in Nome had the story first hand in the winter of 1973, but Joiner's report was in 1966.[40]

Numerous online sourced confirm a Robert Larson's employment with Alaska's Bureau of

Land Management in 1973, but no other reports the alleged Sasquatch killing.

John Steel summarizes our next case as follows:

> In May 1967, in The Dalles in Oregon, several teen-age boys spend their nights hunting Bigfoot. One night, they were moving through the woods, when they came to a tree whose branches hung to the ground. Pushing past them, they found a creature 8-10 feet away that was 7 feet tall crouched down. Dave Churchill blasted it twice in the chest with his 12-gauge shotgun, which knocked it down. It rolled over twice, and then ran off. It broke through a fence, snapping three posts off at the ground. The boys returned the next day to claim their prize, but the footprints disappeared after 80-100 yards, and there was no blood to follow.[41]

Autumn Williams, on Oregonbigfoot.com, adds that the broken fence was made of barbed wire, that the Sasquatch "usually seen" by Churchill and friends was nine feet tall with glowing red eyes, and that it "must have weighed half a ton."[42]

We should note that 12-gauge shotguns are the next size down from the aforementioned 10-gauge, being the caliber normally issued to American law enforcement officers and military combat units. "Buckshot," as suggested by its name, is designed to kill deer and other large animals—or human beings. It comes in various sizes, ranging from "T" (.20-caliber pellets), through numbers 1-4, then "O" through "OOO" (.32 to .36 caliber), on to "Tri-ball" (.52 or .60 caliber). All are devastating at close range, and the odds of even a nine-foot-tall biped surviving two blasts to the chest should be nil.

Robert Lindsay offers another tale of Sasquatch slaying, credited to Peter Byrne. He writes:

> December 1967: Teton National Forest near Jackson Hole, Wyoming. Two college students from Marshalltown, Iowa—Lyle Bingaman and Mike Burton—shot and killed a Bigfoot, thinking it was a bear. They were terrified that they had killed a human being and that they would be prosecuted for murder, so they left it where it was and didn't talk about it for a long time.[43]

In May 1968, another Sasquatch was reportedly shot near Delphi, Indiana. As reported to the GCBRO by an unnamed email correspondent:

> My Great uncle said that the creature was "monkey lookin" and about five foot high. My Aunt and Uncle had just finished eating their breakfast when they noticed the creature approaching. My uncle waited until the creature was about 20 feet from the door then he opened the door and 'gut shot' the animal with his .22 rifle. He said that the animal screamed (not surprising) grabbed its stomach, where shot and ran back into the woods. It ran away on its two legs never dropping to all fours.[44]

On its face, this story sounds more plausible than others we have heard. The .22-caliber cartridge is small, and human-sized targets may easily survive multiple wounds unless a bullet finds a vital organ. That said, we have no hard evidence that the incident ever occurred.

Fast-forward to January 1969 for our next incident, involving law enforcement officers. On New Year's Day, the following item appeared in the *Sullivan* (Missouri) *Independent News*.

Man pumps 9 rounds into "Momo" from point blank range

Washington County Sheriff Pete Floros reported that shortly after 2 a.m. on Thursday morning, he and members of the Sullivan Police Department and the Franklin County Sheriff's Department responded to a call in the Hamilton Hollow area, just south of Meramec State Park...

About a dozen officers spent almost two hours searching through the woods for a wounded "beast" that Arbie Boyer of Route 4, Sullivan, said he shot. Mr. Boyer told the officers he pumped 9 rounds at point blank range from a .22 caliber long rifle semi-automatic pistol into the chest of a seven foot tall hairy animal which he said was within 20 feet of the front door of his home. He showed the officers where the first shots were fired and they found the empty shell casings on the ground and were taken as evidence.

Mr. Boyer said the animal walked upright with its head almost met its shoulders and no visible neck and that the arms hung by its sides almost to where the knees should be. He said it was brown in color and appeared to be covered with matted hair.

After the shots were fired, the animal turned and slowly walked away. Boyer said at that time he ran back inside the cabin/house and returned in time to get off one shot from a 45/70 caliber rifle, which he said hit the beast in the area of its right shoulder.

Officers were unable to find sufficient tracks in the dark of that very cold winter morning and gave up the search shortly after 4 a.m. Boyer was able to draw a rough sketch of the beast and what he described fit the description known in this area of the country as the Missouri "Mo Mo," Bigfoot or Sasquatch.[45]

We have discussed the problem of shooting large game with a .22-caliber weapon, but the .45-70 rifle cartridge is something else entirely—a .45-caliber bullet weighing 26.2 grams, driven by 70 grains of gunpowder. Invented in 1873, it was the standard-issue rifle round for U.S. troops during the Spanish-American war and the subsequent Philippine Insurrection. It is, in short, a certified "man-stopper." As to why it failed with Sasquatch in this instance, Mr. Boyer's marksmanship may not be all he claimed.

The *Sasquatch Tracker* website logs our next report, reading: "1969/01/L—Fort Yukon, AK: Man is attacked and then allegedly kills Sasquatch."[46] From the date, we might infer that the incident occurred in January, but the website lists its source as Bigfoot Encounters, where we read: "Near Ft. Yukon, Alaska, August 1969: Jim Ward is alleged to have shot *at* a large hair-covered man while moose hunting."[47] (Emphasis added.) No other source exists for the supposed attack or slaying.

Dennis Bauer offers up two more alleged incidents from 1969, briefly described as follows:

> 1969-11-00; CA, Calaveras; bigfoot shot 3 times from 30 yards away.
>
> 1969-00-00; NM, San Juan; shepherds shoot a bigfoot, two other bigfoots help wounded bigfoot.[48]

His source for the first case is a dead hyperlink, leading nowhere. Bauer credits the IBS database for the second report, where we glean added information that the shooters were Navajo Indians and the Sasquatch was eight feet tall.[49] It seems no attempt was made to track the wounded creature, or to otherwise confirm the incident.

Closing this chapter, we have three reports vaguely dated from the 1960s, with no further specifics. The first, bizarre to say the least, comes from the IBS database, reading: "CLARK SHERIFF LARRY LUND, 1960s: DEAD BODIES, BLOOD SPATTERS IN HOUSE." The "full report" parenthetically promised no longer exists online.[50] My research for *Sasquatch Down* confirms a Larry Lund residing in Vancouver, Washington (Clark County), but revealed no evidence of his service as county sheriff. Lund *did* address the 22nd Annual Ohio Bigfoot Conference in 2010, noting that he was born in 1947 and served in the U.S. Army during 1966-69.[51] Beyond that, nothing, but since the brief description of this incident apparently refers to slain humans, it is probably irrelevant in any case.

Robert Lindsay returns with our next report, which reads: "1960's, Douglas, Oregon: In the Cascades west of the Umpqua National Forest, a farmer shot a Bigfoot and then somehow managed to take it back to his house, where he left it outside. Other Bigfoots then came that night and retrieved the body."[52] Lindsay credits the IBS as his source, but none of the fourteen Douglas County reports preserved online match the incident described.[53]

Finally, we have Dennis Bauer writing: "1960's winter; CA, Shasta; bf shot with 30.06, tracked thru snow to creek."[54] His source, Bigfoot Encounters, traces the report to an "early 1960s" barroom conversation with one Albert Hamilton, a Native American born in 1900 and sixty-six years old when he related the story (i.e., 1966-67).

We should begin by noting that there is no town called Wildwood in California's Shasta County, though several other counties boast towns of that name. With that in mind, Mr. Hamilton claimed his encounter with Sasquatch began late one night, during a dance at the Wildwood Inn. Correspondent Ben Foster, writing to Bigfoot Encounters from Indiana, explains.

> Albert just happened to glance out one of these [windows] just in time to see something or someone looking back at him from a lower crouched position. He and

several other men walked outside to see what was going on. The Bigfoot stepped off the porch and walked north toward the woods, at the same time another man ran to his truck and returned with a 30.06 rifle which he fired only once. His aim was on and it was believed that the Bigfoot was hit; it let out a terrible scream and ran.

There was about 6-inches of snow on the ground; the Bigfoot left many tracks and a blood trail. The men decided to trail the beast and finish him off. After about 30 minutes the men gathered with guns and flashlights and began their trailing.

They followed the Old Bigfoot for several miles to a small stream of water an offshoot of the Trinity River but this is where they lost him. Albert said this creature was near 6-feet tall, hairy everywhere except the face, palms of its hands and bottom of its feet. It was a reddish brown, and smelled very bad, but that was the last time ole Bigfoot ever went to the dances around there.[55]

Quite understandable, but the report still leaves us frustrated. Garbled geography aside, we have—as usual—no evidence to prove that the incident ever occurred. Bauer clearly has the wrong date, and my attempts to find Ben Foster in Indiana proved fruitless.

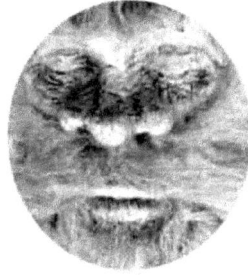

Chapter 4.
Unfriendly Fire (1970-1989)

T he next two decades in our survey include sixty-eight reports of hairy hominids killed or wounded across North America. Robert Lindsay leads the parade with an item simply dated "after 1969," which reads: "Clark, Washington. Near Mt. St. Helens, a man shot and killed a Bigfoot, then tried to sell it, but stopped when he thought it might have been illegal to kill the Bigfoot. No further details."[1] As in other cases, Lindsay cites Ray Crowe's *Track Record* as his source, and while the IBS database does not deliver the "full report" promised for item #2959, it tells us enough: "CLARK SHOOTER Ray Wallace."[2]

Ray Wallace, of course, is the notorious, pathological hoaxer who plagued Sasquatch research from 1958 until his death in 2002, faking footprints, spreading incredible tales of Bigfoot kills and captures, offering nonexistent "genuine" captives, films and photographs to the highest bidder. At his death, relatives tried to cash in on his life of lies, peddling film rights to his story while declaring that Wallace "invented" Bigfoot and that "Bigfoot just died."[3]

In 1977, anthropologists Roderick Sprague and Grover Krantz published a monograph titled *The Scientist Looks at Sasquatch*. In passing, Krantz mentioned that an unnamed Washington hunter had allegedly killed a Sasquatch near Spokane, sometime in 1970. He presented no proof, but the sketchy anecdote has now become an article of faith among some researchers, such as Dennis Bauer[4], and is logged in the IBS database (albeit without the "full report" promised).[5]

Bauer presents the next report, this time of a Sasquatch wounding. He writes: "1970-05-00; CA, Butte; human shots bigfoot 4 times with .22 caliber rifle."[6] Janet and Colin Bord elaborate, pegging the date as May 12 and naming the shooter as Clifford Brush. The shooting occurred, they say, in Butte Creek Canyon near Chico, where Brush had gone to draw water from his well. The wounded creature "went away making cries of pain," and we know nothing more.[7]

Bauer returns to offer a brief account of our next case: "1970 or 1972; FL, Citrus; human shoots a bigfoot."[8] This time, however, he provides a link to the GCBRO's website, where we find an anonymous report from a self-described forty-three-year-old grandmother. Reminiscing of her teenage years, she described weekend outings at a stone quarry where hairy bipeds appeared by night, hooting and lobbing stones at human interlopers. One night in 1970 or '71, while watching one creature climb a nearby cliff—

> I heard a lot of ahs and ohs from our group and then I jumped cause some one shot a gun. Everyone jumped at the blast and Scotty moved the spot light. When he focused back on the ledge the bigfoot was getting up off of one knee and slid against the lime stone wall behind it and started up the wall towards the ravines to the side of the pit. The bigfoot was shot; it looked like he was hit in the side, close to where our belly buttons are. Even the night this idiot shot the bigfoot we still didn't feel threatened or scared. We left early more because of guilt than any thing else.[9]

The tale ends there, with no mention of whether the creature escaped or collapsed at the scene. It is, needless to say, unverified.

Ray Crowe logged our next apparent Sasquatch wounding, dating from the early 1970s. The event reportedly occurred in Chelan County, Washington, where a seven-foot biped attacked a farmer's 150-pound pig in its sty. Crowe was so impressed by the tale that he listed it twice, as #73 and #2119, with somewhat different summaries. The first reads: "BF after pigs; dropped 150# pig from biting neck and drug away; screaming; shot 7' BF; blood." The second tells us: "7 FT BF STEALS PIG, SCREAMS, SHOT BY 10 MEN, BLOOD TRAIL."[10]

Being shot by ten men clearly implies a target riddled with bullets, yet the creature managed to escape. Crowe credits the story to Sasquatch researchers Tim Hurst and Richard "Rip" Lyttle. I attempted to reach both of them, in vain.

Dennis Bauer returns with the next shooting report, which reads: "1970's early mid; LA, Vernon; a bigfoot vocalizes loudly when shot by hunter."[11] He links readers to the GCBRO website, where we find an anonymous report filed in October 1989. The final paragraph of that account tells us:

> My father, who is now age 73 and still resides in Vernon Parish, Burr Ferry area, heard a eerie sound when he went coon hunting in the early to mid 70's near the (Location Withheld). He was coon hunting and shot at some kind of figure that stood about 15 foot tall. He said, when he shot the animal it screamed extremely loud and it sounded like a squeal that he had never encountered before in his lifetime. He also said, "This sound was definitely not a panther squeal."[12]

If this event occurred, in fact, we have no way of telling whether the creature was wounded or simply frightened. The anonymous hunter did not pursue it.

Dennis Bauer gets our hopes up with his next report—"1971 to 1976; FL, Citrus; humans shoot bigfoots"[13]—a veritable massacre spanning five years. Ardor quickly fades, however, when we refer to his source, Rick Berry's *Bigfoot on the East Coast*. In fact, Berry's only mention of Citrus County during the stated time period describes a case from November 1975, wherein witness John Sohl "fired the flash" on his camera at three hairy bipeds.[14] On balance, it seems likely that Bauer garbled the GCBRO report from 1971-72, quoted above.

Bauer's next report is equally frustrating. He writes: "1971-00-00; CA, Los Angeles; bigfoot shot 4 times with a .30-.30 BFF."[15] According to Bauer, "BFF" refers to a now-defunct website, Bigfoot Fact or Fiction, which took its data from the Bords' *Bigfoot Casebook*.[16] There, we read that the shooting allegedly occurred at Littlerock Dam (misspelled in the text as "Little Rock"), near Palmdale in northern Los Angeles County.[17]

In passing, once again, let us note that the .30-30 Winchester rifle cartridge, introduced in 1895, is a popular hunting round capable of killing deer, moose, caribou and bear. Its name refers to a .30-caliber bullet propelled by thirty grains of gunpowder, hurling an eleven-gram projectile down range at 2,227 feet per second.[18] The notion of any hominid surviving four hits from such rounds strains credulity—but as we have seen from countless human gunshot wounds, nearly anything is possible.

Denis Bauer lists three more cases from 1972 on his Bigfoot Shootings website, but the first—"1972-00-00; TN, Morgan; human hits a bigfoot with brick, bigfoot chases human"[19]—clearly does not qualify. The other two are worth pursuing.

> 1972-00-00; FL, Citrus; human shoots a bigfoot BOTEC.
> 1972-06-00; TX, Rusk; bigfoot watches campfire for four minutes, humans shoot bigfoot, bigfoot flees TBRC.[20]

"BOTEC" is Bauer's code for Rick Berry's *Bigfoot on the East Coast*. Berry lists four cases from 1972, only one of those involving gunfire. That item, from the Crystal River area, tells us "An anonymous man shot *at* a hairy, bipedal creature that ran off into the swamp."[21] (Emphasis added.) Bauer infers a wound where none is claimed.

"TBRC" refers to the Texas Bigfoot Research Conservancy, known today as the North American Wood Ape Conservancy. That group's website lists five Sasquatch sightings from Rusk County, Texas. None occurred in 1972, and none involved shooting.[22]

Bauer cites Rick Berry as his source for out next shooting: "1973-06-00; MD, unknown; human shoots a bigfoot."[23] Berry, for his part, offers no source for an account from early June, reading: "An anonymous individual shot at and supposedly hit, a very huge hairy, man-ape creature that was walking on two legs."[24] The key word is *supposedly*.

Our next report, from Kentucky, comes in several versions. Dennis Bauer provides the shortest: "1973 autumn; KY, Clinton County; human repeatedly shoots a bigfoot."[25] He links readers to a Kentucky Bigfoot research website, where we read:

Witness Charlie Stern saw what he said was a 6' tall, dark, haired [sic] covered creature with an ape-human face and a bushy, black tail as it killed one of his animals. He repeatedly fired at the creature from close range to no effect. After many shots the creature at last seemed wounded and ran off on two legs.[26]

Bushy tails are not a feature commonly attributed to Sasquatch. Janet and Colin Bord add that the prowler left nine-toed footprints, saying that it "killed livestock but was itself unaffected by gunfire until wounded by farmer Charlie Stern; sightings of this, another and a youngster then ceased."[27]

Even more confusing is a Pennsylvania case reported from the same time period. Dennis Bauer introduces the muddled event with two separate listings on his Bigfoot Shootings blog.

1973-00-00; PA, Fayette; human shoots at a bigfoot twice. BOTEC
1973-11-00; PA, Fayette; man shoots at a bigfoot twice, second time it grunts and cries.[28]

The first of Bauer's two accounts leads us back to Rick Berry, whose report places the shooting in November. Berry writes: "A man shot at a large, hairy, man-like creature and it disappeared. He saw it again later and shot at the creature. It grunted and cried."[29]

Bauer's second report links readers to the Pennsylvania Bigfoot Society's website, where we read:

October 25, 1973: North Union Township: A man and two boys saw a UFO land in a field. Upon investigating, they observed 2 tall, hairy bi-pedal creatures walking along a fence line. The creatures had matted, dirty brown hair, glowing green eyes, broad shoulders, and small necks. They walked stiff-legged giving them an overall robot-like appearance. The man shot at the creatures, which caused the spherical UFO that had landed in the field, to leave the area immediately. Although the witness felt he didn't miss, the shots had little effect on the strange creatures. They simply turned around and traversed their way back along the fence line. A luminescent ring on the ground remained where the UFO had once been. The witness left the area and returned later with a Pennsylvania State Trooper. Although the glowing ring near the ground was still visible, its intensity had diminished. Several hours later the luminescent ring completely dissipated.[30]

Aside from changing the incident's date, this account's introduction of UFOs automatically places it off-limits for dedicated "flesh-and-blood" cryptozoologists. Whatever any reader's personal opinion on that subject, we may all agree that the report provides no evidence of Sasquatch being shot, much less slain.

Our next case, from Florida in 1974, ranks among the more confused found in print. Dennis Bauer opens by listing *three* separate incidents. Those items read:

1974-00-00; FL, Dade; Police officer shoots at bigfoot, bigfoot runs into swamp SFB.

1974-01-00; FL, Lee; bigfoot kills pony, human shoots at a bigfoot BOTEC.

1974-01-00; FL, Palm Beach; human shoots a bigfoot BOTEC.[31]

Bauer provides an active link to "SFB," but it leads us to the KeutuckyBigfoot.com website's cases from Clinton County, irrelevant to Florida. In private correspondence, Bauer explained that "SFB" was an accidentally "misspelled acronym," meant to read "SAFB"—which, in turn, refers to a now-defunct website called Skunk Ape/Florida's Bigfoot. He graciously provided excerpts from that page, archived in his personal collection, including one that reads:

> 1974—N. Miami, Dade Co, Fl. Hwy 27—Officer Hollemeyal dispatched to accident saw a creature in road in front of him. He said it was bulky and strong looking. He shot at it and it run into swamp. There is also a report of a terrified dog along the road in the same area. The Officer said creature was covered in black hair with a silver or white strip down its back.[32]

I would exclude this item based on the officer merely shooting *at* Sasquatch, but Rick Berry tells a different story in *Bigfoot on the East Coast*. There, he writes: "1974, January 9th, Highway 27 near Fort Lauderdale [Broward County, not Dade], Patrolman Robert Hollemeyal shot at *and hit* a very bulky, hairy, ape-like creature which ran away on two legs."[33] [Emphasis added.] The Bords—and most other published sources—agree with Berry's spelling of Officer Hollemeyal's name. They also clock the creature's speed, by some means unknown, at twenty miles per hour.[34]

Bauer's other two Florida cases, both citing *Bigfoot on the East Coast* as their source, are confused. Berry lists no Lee County incidents from January 1974, but he does report another shooting on January 9, at Immokalee in Collier County, where he says a Sasquatch "killed a pony and jumped a fence. While the creature was running away it was shot at but apparently was not hit." On the same day, in Dade County, motorist Richard Lee Smith reported striking Sasquatch with his car. Another driver reportedly saw the creature limping, some two hours later.[35]

All in all, a bad day for Sasquatch in the Sunshine State—but more confusion lies in store. The Bords—citing an article from *Startling Detective Magazine,* published in March 1976—write that the Hollemeyal shooting and Richard Smith's collision with Sasquatch both occurred in Fort Lauderdale. They omit the Immokalee pony-slaying and near-miss shooting entirely.[36]

Pennsylvania also witnessed violent Bigfoot action during 1974. Dennis Bauer's first report tells us: "1974-00-00; PA, Indiana; human shoots bigfoot seen on porch."[37] His hyperlink takes us to the Bigfoot Encounters website, where we find the following paraphrase of an article published in *Saga* magazine, sometime during 1974.

> At dawn, Roger McCracken saw an 8 foot grayish hair covered giant by his garage. Next evening his stepdaughter, Betty Ruffner saw it. Huge, slouched, arched-back with long arms and tremendous stride. On September 13th, Betty and her teen-aged son, Raymond, saw it at night on the back porch and he shot at it twice. He

thought he hit it in the leg. It leaped the road and ran into the bushes. They heard screams. "Blood" spots tested and found to be mixture of saliva and juice from apples from barrel on porch.[38]

It might be helpful to know who tested the "blood" spots, but in any case, no evidence remains. Bauer reports the second Pennsylvania incident as follows: "1974-02-00; PA, Fayette; humans shoot multiple bigfoots BOTEC."[39] Rick Berry dates the incident from February 6 and writes:

> A woman heard a noise outside her house, went to the doorway with a 16 gauge shotgun, opened the door, and saw a 7 foot tall, ape-like creature in the doorway. She claims to have shot at the creature and it disappeared in a flash of light. Her son who lived nearby heard the shots, went to investigate, and saw several hairy creatures in the area. He shot at them with a pistol.[40]

And missed them, apparently—or, perhaps, they vanished in a "flash of light" as well. The Bords discuss this incident at length in their *Bigfoot Casebook*, calling the female shooter "Mrs. A." They state that she fired at the creature's midriff with her shotgun—the next size down from a 12-gauge—from a range of six feet, whereupon there was "a brilliant flash like a photographer's flashbulb and the creature had disappeared, leaving no trace whatsoever." Her son-in-law, rushing to help her, saw "four or five" hairy giants and fired at them twice, to no effect. Both shooters also reported a red UFO resembling "a Christmas tree decoration" hovering over some trees 500 feet distant.[41]

Dennis Bauer's next report for the year reads: "1974-05-00; NC, unknown; human shoots a bigfoot BOTEC."[42] His source, Rick Berry, offers more details, writing that the incident occurred on "South Mountain"—actually the South Mountains State Park, in Burke County—where an unnamed witness saw Sasquatch and made plaster casts of its footprints on four occasions. At the last encounter "one appeared five feet behind him, [and] he turned and shot the creature. It fell, but was gone when he returned."[43]

Bauer takes us back to Florida with his next report: "1974-09-00; FL, Palm Beach; security guard shoots a bigfoot SFB."[44] Supporting that item, Bauer generously sent not one, but two accounts from the defunct SAFB website, seeming to report the same event, albeit with different dates. Without corrections, they read:

> 1974-Palm Beach County, Fl. Security guard says he shot at and hit a Skunk Ape, which fled.

> 1975-West Palm Beach, Fl. Security guard shot at 7 ft tall, hairy, either hunched over or no neck creature. Gary K., age 24, guarding construction site, nr.W.P.B. said he yelled freeze but it kept coming. He said what he saw made his hair stand on end. he said that he had drew his gun when he yelled for it to stop. It had an awful smell. It made his eyes water. 1974 report also.[45]

The Bords support him after a fashion, saying that a Palm Beach County security guard "fired at and hit Bigfoot, which fled." Their source, in turn, is listed as an untitled article from the *Palm Beach Post* (incorrectly titled the *Palm Beach Post News* in their chronology), published on February 11, 1977.[46] Rick Berry elaborates on the story, dating it from September 24 and saying that the guard "fired six bullets into a 7 foot tall, hairy, hulking creature that kept coming towards him. He thinks he hit the creature because it grabbed its chest and took off running. He said the creature smelled like rotten eggs."[47]

The Bords alone report our next case, from the vicinity of Holly Springs, Arkansas, in mid-October 1974. Their brief account states that several witnesses saw hairy bipeds, while "one man shot at and hit Bigfoot."[48] As usual, the shooter had nothing to show for it.

Dennis Bauer provides our next shooting case: "1974 fall; IA, Woodbury; human shoots and wounds bigfoot."[49] As evidence, he links us to Bigfoot Encounters, where a brief item reads: "1974 near Stone State Park, Sioux City, Iowa: A man claimed to have wounded a bigfoot with a deer rifle."[50] Again, we see how a "claim," this one drawn from a November 1978 issue of the *Des Moines Sunday Register*, is transformed via Internet legerdemain into a statement of "fact."

Florida produces our next report, rendered by Bauer thusly: "1974-11-00; FL, Collier; humans shoot a bigfoot."[51] His link draws us to the GCBRO's website and a report from several anonymous hunters who met Sasquatch while stalking game in Collier County's Corkscrew Swamp. The eight-foot-tall creature "just stood there staring at them. They said it smelled really bad, like an animal, and it scared them, so they shot at it with their shotguns. The thing screamed and fled on two feet into the dense underbrush. They pursued it to no avail, but did find signs of blood. They were so shaken that they broke camp the next morning in fear of retaliation."[52]

Florida confuses us further with a series of reports from February 1975. Dennis Bauer lists them as follows:

> 1975-02-00; FL, Alachua; car knocks down a bigfoot, human shoots bigfoot BOTEC.

> 1975-02-00; FL, Lee; human shoots a bigfoot BOTEC.

> 1975-02-00; FL, Lee; human shoots a bigfoot, then repents BOTEC.[53]

Bigfoot on the East Coast includes all three incidents, though in different order. The Alachua County road rage incident occurred south of Gainesville, on Williston Road, where motorist Steve Voreh allegedly struck Sasquatch with his car, then shot it four times in the chest with a .38 Special revolver. None of those injuries stopped it from fleeing the scene.[54]

Both Lee County shootings involved a single gunman, Richard Davis of Cape Coral. On February 2, he reportedly shot a nine-foot-tall Sasquatch once, "but could not fire again because of the human like appearance of the creature." The following day, he saw it again and forgot his earlier "repentance," shooting the creature again with a .22-caliber pistol. It escaped on both occasions.[55]

September 1975 brings us another case garbled in translation. Dennis Bauer writes:

1975-09-00; KS, Nowata; humans shoot at a bigfoot IBS.
1975-09-00; OK, Nowata; bigfoot shot 5 times with rifle, once with shotgun.[56]

The first listing is erroneous. Kansas has no county called Nowata, though Bauer may be forgiven for this error, since the IBS database *does* describe a shooting, rather awkwardly, as follows: "BF shot at; walked away; human features, Nowata County; Kansas; Sept 2; 1975; two men saw monster like beast in woods not far from Coffeyville [Montgomery County]. Clifford Bentson; and Marion Parret woke to sound of weird noises; and saw Bf. Unloaded rifle at it and it just walked away. Hairy apelike monster with human features. Seen at midnight close to house; stood right by fence. Unloaded 30.30 at it." The IBS credits its account of the shooting to *Bigfoot All Over the Country,* a book published by Marian T. Place in 1978.[57]

Meanwhile, from Nowata County, *Oklahoma,* Bauer provides a link to the Bigfoot Encounters website. There, we read of an event occurring on September 4, 1975.

> Several witnesses saw a 6 foot creature, hair covered except around the eyes. Three men got best look at it, one of whom *shot at it* 5 times with his rifle. Another man *shot at it* with a shotgun loaded with birdshot. The CBC Radio Show Interviewed Deputy Sheriff Bob Arnold who confirmed the story saying there was a rotten odor too.[58] [Emphasis added.]

Neither marksman claimed to have hit his target, and the creature apparently fled with no sign of any serious injury. We do not know the range from which they fired, but birdshot—while potentially lethal to man-sized targets at close range—rarely inflicts serious damage from a distance.

More confusion awaits us in Pennsylvania, courtesy of Dennis Bauer's website.

1975-10-00; PA, Fayette; human shoots at two bigfoots BOTEC.
1975-10-00; PA, Washington; humans shoot at a bigfoot BOTEC.[59]

While Bauer cites *Bigfoot on the East Coast* as his source for both events, Berry's book only lists one incident for October 1974. It reportedly occurred on October 9, near Monongahela, in Washington County. On that day, three gunmen—including a Marine Corps sharpshooter—saw two large Sasquatches together, and "all three men shot at them with no effect."[60] It seems Bauer has split one incident into two, transferring one to a different locale.

As a native of Kern County, California, I was particularly interested in Bauer's next case: "1976-00-00; CA, Kern; man shoots a bigfoot in chest ten times with .22 rifle IBS."[61] Rushing to the IBS database online, I found Report #1782, repeating claims made by correspondent Mark Burkey in a letter to Peter Byrne. The report says:

> Mark Burkey, Notes from letter of Oct 13, 1976: We have a lot of trees, mountains, and brushy areas (around Bodfish, CA). It all happened one night when a guy came home around 11:00 pm. He drove up to his house and looked around, as he thought

he hit a skunk, and walked over to the garbage can and there it was. He says it was about 8 ft. tall and around 500#. It walked up right over to a girl's house where she screamed and the guy ran over there with a .22 rifle and shot about 10 shots into his chest. He said it didn't even hurt him. He said it wasn't a bear or anything like that, because the forest (service) moved almost all the bears into the mountains.[62]

So close, and yet so far away! The witness remains anonymous, and the second-hand account produced no evidence.

Dennis Bauer's next case reads: "1976-00-00; FL, Marion; humans shoot bigfoot with .22 caliber rifle SFB [*sic*]."[63] In response to my inquiry, he provided the following (uncorrected) item from the defunct SAFB website: "1976—Dunnellon, Marion, Fl. Past five years elderly man claims to have seen a large, hair covered creature. Noted green eyes. Found 5 toed tracks on several occassions [*sic*]. Neighbors said that they saw it and also *shot at it* with a 22 rifle."[64] [Emphasis added.]

 Bauer cites *Bigfoot on the East Coast* as the source for his next incident from 1976, summarized as: "1976-06-00; MD, Baltimore; humans capture or kill possible bigfoot BOTEC."[65] Rick Berry fleshes out that incident.

> Wade Bowere, owner of the Harewood Foot Mart [*sic*], kept track of the dozens of Harewood Park encounters, and claims that in one incident dozens of soldiers from the Aberdeen Proving Grounds entered a swamp and came out carrying some type of huge creature in a bag.[66]

We shall discuss alleged conspiracies to hide Sasquatch remains in Chapter 11. For now, suffice it to say that grocer Bowere proved untraceable during my research for *Sasquatch Down*.

Another Sasquatch slaying—this time, of two specimens—allegedly occurred in Texas, during early 1976.[67] I mention it here to preserve our chronology, but will discuss it fully in Chapter 5, since the claim was not aired publicly until 1996.

Dennis Bauer returns to the Sunshine State for his next reported case: "1976-07-00; FL, Citrus; a bigfoot rolling in the sand by a yard gate is shot twice with a .22 and walks away howling BOTEC."[68] Rick Berry, once again, has a different take on the incident.

> 1976, July 17 and 18th, Dunellon. Donald Duncan's five year old son saw a tall, hairy ape-like creature standing beside a tree in their yard. That night they heard dogs barking, and upon investigating saw the creature rolling in the sand just outside the gate with three dogs. Donald *fired at* the creature twice with a .22, the dogs fled and the creature walked off into the woods "howling like a wolf." He later found one of the dogs with its neck broken and its stomach slit open.[69] [Emphasis added.]

We may only speculate as to whether the Sasquatch was wounded, assuming the incident ever occurred.

Dennis Bauer's next report is vaguely dated: "1976 or 1977; OR, Linn; man shoots bigfoot, bigfoot tears apart man, heavily armed USFS helicopter evac IBS."[70] Examination of the IBS sightings database online reveals thirty-four reports Linn County, Oregon, none of which refer in any way to a Sasquatch being shot, killing a man, or being chased by agents of the U.S. Forest Service.[71]

Bauer's first confirmed account from 1977 reads: "1977-00-00; IN, Dearborn; human shoots at a bigfoot 12 times with a .22 caliber."[72] His hyperlink connects to Bigfoot Encounters, where we see:

> Dearborn County, Aurora, Indiana
> Published in the Cincinnati, Ohio Post April 20, 1977
> Tom and Connie Courter were coming home at night...as the husband got out of the car, a monster collided with the car. The next night when they came home they saw it perched on the hill. Tom Courter *shot at it* 15 times with a .22, describing it as a hairy, apelike creature approximately 12 feet tall. It made a noise like "ugh."[73]
> [Emphasis added.]

Assuming Courter was lucky with one or more of his fifteen shots, we see once more the futility of using .22-caliber weapons against big game.

Dennis Bauer finishes 1977 with five more summarized reports from four states.

> 1977-05-00; NJ, Sussex; humans shoot a bigfoot.
> 1977-08-00; NJ, Sussex; human shoots a bigfoot, bigfoot retaliates.
> 1977-10-00; FL, Broward; bigfoot tears human's shirt, human shoots bigfoot BOTEC.
>
> 1977-10-00; PA, Westmoreland; human shoots a bigfoot BOTEC.
> 1977-11-00; FL, Marion; human shoots a bigfoot BOTEC.[74]

Bauer's hyperlink for the first case takes us to the New Jersey Bigfoot Reporting Center's website, where we find "one of the most famous cases in New Jersey" reported as follows, from Wantage on May 12: "Bigfoot visits the Sites farm in Wantage and kills some rabbits, raises havock [*sic*]. Returned later and *was shot at* by family." (Emphasis added.) A newspaper article reproduced from the *New Jersey Herald* (May 17) claims four shooters "opened up" on the creature, firing more than thirty rounds from two shotguns and two .22-caliber rifles, but no claims of hits were reported.[75]

The August case links Bauer's readers to the GCBRO's website, reporting an incident from August 15. That report describes a man firing a pistol at a tall, dark-haired biped which fled the scene, apparently unharmed, but returned the following night, attacking a pig named Wilbur in the barnyard, leaving it with a "pretty deep" scratch on its face.[76]

Bauer's next three reports all come from Rick Berry. The first refers to an encounter reported by security guard Donnie Hall, who met Sasquatch while guarding an Opa-Locka nursery (Dade County, not Broward) on October 3. The prowler allegedly ripped off Hall's shirt, after which he "fired several shots at the creature but is not sure whether he hit it."[77]

That same month, in Latrobe, Pennsylvania, an unnamed man came home one night to find his dog dead, hanged by its neck, with an eight-foot-tall, red-eyed monster standing nearby. Berry says, "He shot at the creature, but the bullet bounced off with no effect."[78]

Back in Florida's Ocala National Forest, one month later, Berry writes: "A hunter fired six rounds at an ape-like creature weighing approximately 800 lbs."[79] Again, the report included no claims of actual hits.

Bauer opens the new year with two more sketchy reports, from Washington and Pennsylvania.

>   1978-00-00; WA, Pierce; soldier shoots a bigfoot.

>   1978-04-00; PA, unknown; human shoots a bigfoot BOTEC.[80]

For the first incident, Bauer provides a link to Bigfoot Encounters, where we find the story of Edwin Godoy, a soldier standing guard overnight at Fort Lewis, near Tacoma. Around 12:15 a.m. he saw a "very big, huge" figure covered in long gray hair approaching his duty post. Godoy said, "That thing started running towards me, so I shouted a halt three times, asking that thing to stop and identify itself. As it wouldn't reply I made a first shot to the air and then I shot at him or 'it,' I don't know how to call it. The hairy thing grabbed its chest and emitted a loud moan, stopped and then ran to his right, disappearing into the forest." Investigators examined the prowler's footprints and collected samples of "blood that looked red, but strangely oily and fresh looking." Officers then allegedly embarked on a cover-up discussed more fully in Chapter 11.[81]

Janet and Colin Bord report our next case, from the vicinity of Owensboro, Kentucky—in fact, they list it twice in their *Bigfoot Casebook* chronology. The first version, dated mid-August 1978, says "Several men cornered Bigfoot beside pond and shot it with pistol from 10 ft. It fled into woods, leaving no trace of blood."[82] The second version, separated from the first by seven other sightings from around the country, reads: "Larry Nelson, brother, & two friends fired three .45 bullets into Bigfoot's chest from 45 ft.; it ran away into woods apparently unhurt."[83] Discrepancies in range aside, the two reports appear too similar to fit separate incidents. That impression is enhanced by reference citations crediting both accounts to a single article, written by one Keith Lawrence for a long-defunct UFO journal in May 1979.

Dennis Bauer closes out 1978 with two more reports of Sasquatch shootings.

>   1978-10-00; OR, Columbia; bigfoot shot between the eyes, rolls off road, shooters flee IBS.
>   1978-00-00; PA, Fayette; bigfoot smashes windshield, human shoots bigfoot BOTEC.[84]

A review of the IBS sightings database reveals nineteen cases reported from Oregon's Columbia County, none of which include a shooting as described by Bauer.[85] Rick Berry places the second event somewhere south of Uniontown, Pennsylvania. According to him:

> A man driving to work on a sleety morning saw a tall, hairy creature sitting by the side of the road. The creature got up and smashed the man's windshield with its fist. The man grabbed his pistol and fired two shots at the creature. It screamed and ran into the woods. Police followed a trail of blood that led to an old mine shaft just across the West Virginia line. They didn't go in.[86]

Perhaps coincidentally, Berry's next report, also undated, describes two children spotting another Sasquatch—or the same one?—south of Uniontown, telling adults that the creature "appeared to be limping."[87]

One of Dennis Bauer's more confusing items reads: "1979-06-00, 1980-03 & 04; KY, Boone; human shoots a bigfoot with shotgun."[88] As written, that implies three shootings, but the hyperlink provided takes us to the BFRO's website, where we find a rather different story. That report, lifted verbatim from a June 1980 issue of the now-defunct *Creature Chronicles* newsletter, features no shootings from June 1979 or April 1980. The single incident involving gunfire—by Dave Stulz, on March 31, 1980—was a clean miss, the article clearly stating, "No definite physical evidence was ever found other than the spent bullets from Dave's shotgun that were found lodged in a large tree down by the river."[89]

Our last 1970s report with a date attached comes, again, from Dennis Bauer: "1979-08; WV, Mercer; bigfoot looks into kitchen window, human shoots the bigfoot."[90] Bauer links his readers to another report from the BFRO, which supports his account—to a point. Drawn from the memory of an 11-year-old boy, three decades after the fact, the incident reportedly occurred on August 10, eight miles from Princeton, West Virginia. After the window-peeping incident, four armed men left the rural home and fired a volley of shots, then returned "pretty well freaked out."[91] Again, there is no claim that any shooter struck his moving target.

A final shooting from the 1970s comes to us with no discernible date. Posted on the Sasquatch Tracker website, it reads: "1970s/06/S—Yakutat, AK: Report of berry pickers shooting 7 foot sasquatch."[92] According to the site's symbol key, "06" denotes the sixth report from the year(s) listed, while "S" indicates an eyewitness sighting. The listed source, IBS Report #3597, is accessible online, but it presents only a third-hand tale of an attempted shooting: "Yakutat man, Steve Johnson, reported to Fred Bradshaw, berry pickers *shot at* seven-foot sasquatch, the creature returned in following years."[93] (Emphasis added.) This claim of a Sasquatch shooting, like so many others, winds up as a miss.

Awful Eighties
Dennis Bauer opens a new decade with this item: "1980; OH, Vinton; human shoots a bigfoot, bigfoots throw boulders at trailers IBS."[94] The IBS database credits researcher Peggy Tillman for the report, summarized as follows:

> At the Wayne Nat[ional] Forest in 1980 a boy shot and wounded a creature, and

blood was found at the site. It was after this shooting that all hell broke loose. Large boulders were thrown at the trailers where people lived. This scared some of the residents so bad that they moved out of their home. Plaster casts were made of tracks that were found.[95]

Dennis Bauer's next report reads: "1980-03-31; KY, Boone; a man shotguns a bigfoot in his yard and it jumps in the river, Sheriff's divers investigate."[96] This is, in fact, a duplication of an incident discussed above, linked to the same *Creature Chronicles* article previously cited.[97] Bauer reports a third shooting from 1980 as follows: "1980-10-00; KY, Fleming; bigfoot raids freezer, human shoots bigfoot."[98] He provides a link to the BFRO's website, where we find the tale of J. L. Tumey.

Mr. Tumey was sitting home watching the baseball game when he heard a strange noise originating from the back porch. Thinking that a prowler was present, he retrieved his pistol and went around back to investigate. He saw what looked like a large dark man-like entity running towards the woods. He *fired shots at it*, then ran back to his trailer for more bullets. He then saw it again by an old stable. *He fired at it again, but missed.* Tumey discovered that the freezer on the back porch had been broken into and some frozen chicken was removed and scattered outside. He also found some unusual white hair and impressions on the ground.[99] [Emphasis added.]

Not a Sasquatch shooting after all, therefore, but only an attempt to bring the creature down.

Bauer's next listing reads: "1980 fall; AL, Lauderdale; a screaming bigfoot throws a hunter down a hill, the hunter shoots the bigfoot which runs away."[100] A hyperlink to the BFRO's website presents a second-hand report from an anonymous email correspondent, relaying an event described by an unnamed cousin. The crucial portion reads:

He stated he then heard a grunting sound and limbs snapping off trees. He sat real still until the creature stepped right out in front of him. He described this creature as at least eight feet tall, and very ape like. He states the ape screamed at him in a blood chilling tone, out of fear he states he fired in the air. He states the he tried to run, but the ape grabbed him and slung him down a ridge. He says he recovered and fired his rifle at the ape as it charged him. He states the ape then fled, as did he and the scared dog.[101]

We have no evidence the frightened marksman's aim was true.

Bauer provides our only case from 1981: "1981-09-00; NC, Cleveland; human shoots a bigfoot BOTEC."[102] Rick Berry pegs the date as September 9, writing that Shelby resident Robert Hunt met Sasquatch on Wells Road, while out for a stroll. "He ran for his gun and fired at the creature. His mother Jean Hunt also got a glimpse of the creature. She said it was dragging its leg."[103]

Wounded? Perhaps.

Dennis Bauer furnishes all four of the available reports from 1982.

> 1982 spring; MD, Carroll; human shoots a bigfoot BOTEC.
> 1982 fall; TX, Cherokee; human shoots a bigfoot with shotgun, 3x, and .357 5X
> TBRC.
>
> 1982-00-00; GA, Macon; human shoots a bigfoot thru bathroom wall.
> 1982-00; OR, Columbia; fisherman shoots bigfoot, follows blood trail until losing
> trail IBS.[104]

Rick Berry places the first incident near Sykesville, where, he writes: "Williard McIntyre saw and *shot at* a hulky, mangy ape-like creature."[105] The story claims no hits on Sasquatch.

While providing no link, Bauer's second item apparently refers to the North American Wood Ape Conservancy's website. That site presently lists two Sasquatch sightings from Cherokee County, neither resembling the shootout Bauer describes.[106]

Georgia's case includes a link to Bigfoot Encounters, recalling an adult Florida resident's memory of an incident occurring at Oglethorpe, Georgia, when she was seven years old. In that account, a six-foot hairy biped was lurking around the witness's home. "Her stepfather grabbed his gun and was looking out the bathroom window when it came up close to the back of the house. Her stepfather shot at it through the bathroom wall. He hit the Bigfoot, but it ran off into the woods."[107] The tale offers no explanation of *how* they knew the fleeing creature was wounded.

Finally, we follow Bauer's reference citation to the IBS database online, where nineteen reports are listed from Columbia County, Oregon. None includes any reference to a shooting or blood trail.[108]

Bauer provides our only case from 1983, summarized thus: "1983-10-00; OK, Carter; human shoots a bigfoot, bigfoot wrecks house interior."[109] Thankfully, his link to the GCBRO's website offers further details, including a date of October 13. It refers to the creature in question as "MoMo," a nickname normally reserved for Sasquatch in Missouri, as it stands for "Missouri Monster." The full report, seemingly written by a law enforcement officer from Wilson, Oklahoma, reads (uncorrected) as follows:

> MoMo coming through a door he just ripped off it's hinges. Loud roaring and
> thumping on back outside wall of house. Parents watching television, children in
> bed. The father in this report was a friend of mine even though he was and is sort of
> a bad apple. Would prefer to keep his name private. He's currently in prison and his
> family probably doesn't need the attention. However, the report should be made
> public due to it's potential to get others aside from me to bring out some more info.
> The father is an avid hunter and probably angered bigfoot hence bigfoot came to
> his home [which was isolated] and went around the house knocking out windows
> and beating on the walls. The father got down his 12 guage and fired a shot through
> the kitchen window at a shadow as it moved toward the back door. A loud yowl was

let out, leading the father to believe he had hit the bigfoot. Shortly thereafter the screen door and door to the back porch were torn off their hinges. The father knocked over the refrigerator and lodged it against a substantial back door. He heard the creature trying to tear the back door off it's hinges as he was getting his family to the pickup truck. The father ran back into the house to retrieve his shotgun and left it due to the bigfoot being half way in the kitchen (he tore the door in half length ways, the door had steel panels.) The family made it into town and called the Police Department. I and three of my deputies went back out to the house to investigate the report shortly after the father called in the report. I had known the father for a long time, and even though he was known to throw a punch every now and then he was no liar. What my deputies and I found was a house that was so damaged that it looked like a twister had gone through part of it. Every wall in the house had been demolished, the kitchen and bathroom facilities torn out, and the only reason I could figure the bigfoot didn't completely destroy the house was that there was evidence that it had probably nearly electrocuted itself by biting the back of the television. The fathers shotgun was an 870 Remington and it looked like a pretzel instead of a firearm. The next morning we went back out there and the place looked even worse in the daylight.[110]

Next up, from 1985, Dennis Bauer writes: "1985; OK, Payne; humans shoot a bigfoot."[111] Another link to the GCBRO offers a brief second-hand account of a tale told by an anonymous witness.

About 1985 he and some other guys were partying by the Cimarron river near Ripley whenever they saw a bigfoot. Whenever it realized they were all looking at it it ran off through the brush and left a large hole in the brush where it crashed through. The guys returned later with guns to look for it and saw it again, whereupon they fired at and hit it. It yelled and ran off through the brush.[112]

Lacking any claim of blood spoor or other physical evidence, we may only speculate as to whether or not this might be a hunter's version of the cliché fisherman's story, regarding "one that got away."

Leap forward to May 1987, for Bauer's report reading: "1987-05-00; AZ, Maricopa; human shotguns a bigfoot."[113] Another link to the GCBRO graces us with a report from an anonymous witness whose strange style of writing makes it rather cumbersome.

This was a Night with a Full Moon so besides Our Coleman Lantern You could See Pretty Good. As I was laying there Half Asleep, It all the sudden Got Dark and it Awakened Me only to See This Huge Hairy Thing Not 10 Feet Above Me!!!!!!!! I Shiver to this Day Just Writng You About It. All I Said was "Oh Great A Damn Bigfoot" I Thought I was Dreaming But I Was Not. There It Was Right Above My Head!!!!!!!!!!!!!! I Got To My FEET and it took off into the Darkness. I Cannot Be Sure But I think It was a Female because It had what appeared to be Mammories. One Of My Friends was just outside Camp taking a Relief and I Yelled to Him "Look Out " and He came Running back to Camp Scared to Death. Evidently He Ran Straight into It coming Back To Camp. By This Time We were All Wide Awake and my other Three Friends were Armed with Nothing but Hunting Knives and We all Stood Gaurd in All Directions, My Other Friend was so Scared He Was Useless to Us. We were Boxed In and No Way Out Till Daylight. Then Came that God Awful Yell and We were Scared out of Our Skins. The Rumblings In The Bushes returned and Then We Heard The Fish in The Slew Going Crazy And We Found Out Why. Not 15 Feet from Us Stood This Huge Creature In The Middle Of The Water!!!!!!!!!! This Thing Stood About 9 Feet Tall I Know because I stand 6 Feet Tall and In That Water it Came Right Up To My Neck but On this Creature, It Barely came up to It's Waste. We were Terrified as It stood there Looking At Us. I did Not Want To Shoot God Help Me Because I Was Not Sure, It Looked Almost Human!!!!!!!! I was keeping My Rifle Pointed at It just in Case. Then Our Lantern Started Running Low on Fuel And We Had No Backup except Our Fire. We Soon Found The Bright Lantern Light Was The Only Thing Keeping It Back. Lantern Low On Fuel, We Decided to build the Fire As High As We Could And Refuel The Lantern Fast. My Friends did so and I held My Shotgun on It. Just as the Lantern went out; It Charged Straight For Me So I Fired At Point Blank Range Dead In This Things Chest But It Did Not Go Down!!!!!!!!!!!!!!! My Friends got the Lantern going and There It stood Holding The Right Side Of It"s Chest. It just Looked at Us and Then It Turned and It Cut Through The Waste High Water On It 25 Feet to the Opposite Shore in Seconds and Crashed through the Trees and It Started Yelling Again.[114]

This tale smacks of poorly written fiction. At the very least, no experienced shooter would call his shotgun a rifle, or vice versa.

The remainder of our cases from the 1980s also come from Dennis Bauer, only one with a specific year attached: "1988 fall; VA, Roanoke; campers shoot a bigfoot 6 times with shotgun buckshot."[115] Bauer's hyperlink leads us to the Virginia Bigfoot Research Organization's website and a sighting by witness "Bryan," from the vicinity of Mount Pleasant.

It was our third night in the woods and we had just got in the tent to retire for the night. We heard a loud call that sounded like a girl screaming that faded into almost an owl sound. About 5 min. later we heard it again and it was getting closer to us, it was a sound we have never heard before so we decided to get out of the tent. We stood around the campfire and we heard it again but this time it was real close and we were getting a little scared so we picked up our shotguns and waited. We could hear it walking along the ridge above us, it stopped for a second or two and then we could hear it running down into the valley at us, we both raised our guns and when it came into site we both emptied our guns at it, it let out a scream and ran very fast back up the mountain. We were very scared and decided to get off the mountain. We went back the next morning to get all our gear and the tent had been knocked down and all our food and clothes were scattered everywhere but none of the food had been taken. We went to look for any blood left behind but could not find any. We fired a total of six rounds of buckshot, I really don't know how we missed, it was only about 20 yards from us.[116]

Three shootings remain for the 1980s, none with specific dates attached. Dennis Bauer writes:

1980's mid; AR, Johnson; two separate shootings of bigfoots.

1980's late or 1990's early; OK, Atoka; a bigfoot flees when shot with a 30.06 rifle.[117]

On the first report, Bauer's hyperlink takes us back to the North American Wood Ape Conservancy's website. That report, from autumn 1985, involves no shooting, but comments appended by investigator Tal Branco read as follows:

> This witness and many other people who reside in this area of the Ozarks independently relate anecdotal accounts of these types of encounters. One such incident reportedly occurred in a settled hollow near Haggarville many years ago. In that incident a farmer was aware of a large, hirsute bipedal animal that removed chickens, pigs and calves from his property, but he was initially afraid to try to stop it from stealing the animals. The subject continued to steal the man's livestock, but would take no more than half the animals the man owned when autumn arrived. The man reportedly got fed up with the thefts, and lay in wait for the subject one night. When it appeared, he supposedly shot it with a shotgun. The subject was not killed on the spot, but, according to the story, never returned to the farm.
>
> Another anecdote from recent years involved two men who were poaching deer at night with the use of headlights in one of the local gas fields. They turned their vehicle around at a gas well pad at the end of the road. While turning around, their spotlight revealed a figure they later described as a "bigfoot." The "shooter" fired at the figure and reportedly wounded it, as drops of blood were found the next day.[118]

Bauer's final report links back to the BFRO's website, more specifically to a report clearly dated June 1992. That report involves no shooting, but does include this remark from the anonymous correspondent: "My friend told me that one of his uncles supposedly *shot at* the thing with a rifle while out hunting in those woods. That event had happened several years before."[119] (Emphasis added.) Again, leaving aside whether the incident occurred at all, there is no claim of Sasquatch being killed or wounded.

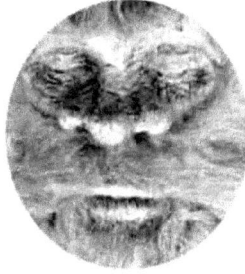

Chapter 5.
Muzzle Flashes and Flashbacks (1990-2014)

O ur final chapter on North American Sasquatch slayings spans twenty-four years and includes 52 reports of manimals struck by gunfire. Eighteen of those claims are reported with no dates attached, and are collected here as a matter of convenience.

We owe our first seven cases from the 1990s to indefatigable Dennis Bauer and his Bigfoot Shootings website. His first case, and the only one on tap for 1990, reads: "1990-05-00; TX, Cooke; human shoots a bigfoot."[1] A hyperlink delivers us to the BFRO's website, for an anonymous account of young campers meeting Sasquatch near Ray Roberts Lake. After one nocturnal sighting—

> The next night we went back with more guns...We went back to the same place and waited. At the same time as the night before we heard the roar again. This time it was further away, then it screamed again yet only a few feet from us. We got the spotlight and lit up the area and that's when we saw it running from us. It went from sounding far away to ten feet with out making a sound in the heavy brush. When we saw the creature, it was about eight feet tall and stood straight up with its back to us. Then as it ran away it stopped as if it was looking at us. Then Russ shot at it with the shotgun. I thought it hit it square in the chest. Then it made the same roar only more intense. Then it pushed a cottonwood tree over... We left and did not return until the following summer.[2]

Bauer's next report is the only one on file for 1991. It reads: "1991-00; IN, withheld; hunters shotguns a bigfoot in the eyes and the bigfoot chases them through the woods."[3] Its source, the GCBRO's website, presents a second-hand report from an anonymous "friend of Timmy," an otherwise unidentified hunter, and immediately raises a problem with dating. While the report was submitted in 1991, and listed by Bauer as a shooting from that year, the text clearly states that the incident occurred "seven or eight years ago"—i.e., in 1983 or '84. From that shaky beginning, we go on to read:

Timmy...was hunting at night with a friend. They were deep in the woods and were using a caller. It sounds like a wounded rabbit...After several calls, he and his friend waited in the dark and heard a couple of cracks. Timmy panned the darkness looking for a set of eyes to shoot at. At a brush pile, he saw a set of eyes about 8 to 9 feet off the ground. At first he thought that an animal was sitting on the brush pile. He then noticed the eyes were wide and set apart by about 5 or 6 inches. He aimed the shotgun and fired! The animal let out a blood curdling scream! Timmy and his friend took off running through the darkness with something hot on their trail...After about a half mile to a mile, the animal stopped and the hunters left the forrest [sic]. The next day, Timmy and his friend went back to the same site in daylight to investigate. At the same location they found multiple prints that a human would make only larger. All around the brush pile and through the woods they found sprinkles of blood where he had shot the animal.[4]

Bauer's next case is indefinitely dated, but is listed here with other 1990s accounts, as it appears on his website: "1992 prior to; OK, Atoka; a bigfoot flees when shoots it with a 30.06."[5] Tracing its source to the North American Wood Ape Conservancy's website, we find a sighting dated June 2, 1992, which involves no shooting but includes a postscript that reads: "My friend told me that one of his uncles supposedly *shot at the thing* with a rifle while out hunting in those woods. That event had happened several years before."[6] (Emphasis added.) As in other cases listed previously, the unnamed shooter claimed no kill.

Bauer presents our sole report from 1992 as follows: "1992-08-00; TN, Cumberland; a bigfoot starts pulling a camper from a pickup truck, the camper shoots the bigfoot."[7] The GCBRO's original report places the incident between Crossville and Westel (misspelled "Westal"), reported by one of a squirrel hunting team. Uncorrected, it reads:

At 3:30am, I woke up to a gun shot! The next thing I heard was my dad yelling for me to get up, something was in the camp! At first I thought he was dreaming but then I realized that he had loaded a gun and fired. I yelled from my tent and asked him what it was. He told me that he was sound asleep when he felt himself being slowly pulled out of the back of the truck. He sat straight up when he saw a hugh dark figure standing at the back of the truck. He thought it was one of us. He said, what are you doing? It didn't say anything. It just stood there. Then it walked past the back of the truck up the drivers side toward the front of the truck. As it went toward the front of the truck, it hit the drivers side mirror bending the mirror frame! He said he didn't know who or what it was so he started fumbling around for a shell for the shotgun. He found a 20 gauge shell and put it in the gun. All of a sudden, he saw it's shadow walking back down the side of the campertop toward the back of the truck. It walked past the end of the truck up to the tent. He said you better get back to bed, thinking that maybe it possibly was one of us. It walked past the front of the tent and that's when he realized it wasn't one of us. He raised the gun and fired. It jumped off to the side of the tent. That's when I woke up and heard my dad yellng for me...What was strange was that we know he hit it from 25 ft. away with a 20 gauge shotgun, but it didn't make a sound.[8]

Dennis Bauer's two cases from 1993 both come from the South, land of quick tempers and triggers.

> 1993-07-00; TX, Titus; homeowner shoots a bigfoot walking along pond in moonlight.

> 1993-10-00; AL, Dallas; a bigfoot follows a hunter and the hunter shoots 33 times towards the bigfoot.[9]

Bauer's links for both reports—the first to Robert Murdock's Sasquatch Information Society; the second to alabamabigfoot.com—were dead in June 2014.

A single case exists for 1994, summarized by Bauer as follows: "1994-12-00; OR, Deschutes; hunters shoot and wound a bigfoot in leg and follow the blood trail for several miles IBS."[10] Consulting the IBS database online, we find that the incident allegedly occurred on December 29, on Mount Washington, west of Sisters. As described by one of the unnamed climbers:

> I saw a big animal that looked like a man but weighed over 600-700 pounds. I could smell it from 40 feet away...we tried to shoot at him; but I was so nervous and scared that I could not hit a damn thing. My friend also had a hunting rifle and he also shot him in the leg; and there was a blood trail for at least 3-4 miles.[11]

Where and how they lost the crimson trail is unexplained.

On June 18, 1995, Regional Emmy award-winning documentary filmmaker Linda Moulton Howe appeared on Art Bell's *Dreamland* radio program, discussing various Fortean topics. One case she mentioned involved an unnamed hunter in Cascade County, Montana, who met an eight-foot, red-eyed hairy biped on some unspecified earlier date. The hunter fired his rifle at the creature, whereupon it vanished "in a flash of light."[12] Howe suggested that the beast might be a supernatural shapeshifter, but the incident reminds us of one reported from Pennsylvania in February 1974. (See Chapter 4)

Bugs Meets Bigfoot
Another Art Bell radio program, *Coast to Coast AM,* produced our next sensational claim of a Sasquatch slaying on April 18, 1996. Ray Crowe was Bell's guest, but a caller known only as "Bugs" stole the show. His story, phoned in to the studio from parts unknown, was riveting. Bottom line: Bugs claimed that he and two hunting companions had slain a pair of hairy bipeds in the Texas Panhandle, sometime during the 1970s. Aging now, and hazy on the dates—some versions claim the incident occurred in January 1976, others pushed it back in time to December 1973—Bugs still recalled the pertinent details.

On the hunting trip in question, he told Bell and Crowe, Bugs and his two amigos had been hunting at night when they glimpsed eyeshine reflecting their headlights from 100 yards. Expecting a deer, a hunter dubbed "Bird Dog" had switched on a rack of high-powered spotlights and fired his .300 Weatherby Magnum rifle at the creature, while Bugs fired his own .243-caliber weapon. Their first shots took the target down, but it rose again. "We fired again at it," Bugs said, "then he jumped up again and we fired a third time and it lunged and hit a fence, but got down into a riverbed and was gone." The hour was late—2:30 a.m.—so they waited for dawn to pursue the wounded beast.[13]

At daybreak they followed a blood trail to the edge of a dense thicket, where it vanished. Bugs lost the coin toss to determine who would brave the thorns to find their kill, crawling fifty feet through thick brush on his hands and knees, until he suddenly confronted a simian female creature, shrieking in rage. Bugs continues:

> The sound was a scream like nothing I ever heard, as she kept coming at me. The hair on the back of my head went up...I had a .44 magnum hand gun with 240 grain shells with an overcharge, and I fired. She was knocked back three or four feet...and again she came at me, and again I fired, and again a third time, all three shots being head shots. This time she stood up, and I fired a fourth time, and at the same time both of my companions could see her as she stood up to 5 or six feet in the brush and they both fired at her also. That time she went down and I could see that she wasn't breathing...and we crawled in. The dead male laid about ten feet away. I noticed they both had [sex] organs like humans. The three of us drug the bodies out of the brush. The bodies looked like they were human, except they were covered with brownish red hair. There was no clothing. The nose was like a human's though the mouth was apelike (no lips). The eyes were half human, half ape, and they had large, protruding, foreheads, with a short neck in front. The overall face was half ape and half human. From the back side there was no neck at all, it was all muscle. The male stood about 7½ feet at about 350 pounds, and the female at 7 foot and about 300 pounds.[14]

Confronted with the corpses of their victims, Bugs and company suddenly feared that "some D.A. out to make a name for himself, win or lose, would have everything to gain with a murder trial." They snapped ten or twelve Polaroid photos (later lost, Bugs said, in a house fire), then buried the bodies somewhere near Elm Creek, Texas. All three gunmen kept the story to themselves until 1993, when Bugs finally told his wife. Hearing Ray Crowe on *Coast to Coast AM* compelled him to phone in and confess, albeit while remaining anonymous.[15]

That call produced dramatic reactions, including a fax from Peter Byrne, vowing to cover any legal expenses Bugs might incur by coming forward, together with an unspecified "monetary reward for an exceptional and unique contribution to science."[16] Still Bugs balked, although he did phone Bell again—on June 5, 2001—this time with Sasquatch researcher Robert W. Morgan in the studio. Grudgingly, Bugs agreed to send Morgan a map revealing where the two creatures were buried, but it never arrived.[17]

The story took another bizarre twist in January 2009, when Loren Coleman wrote online, "Now, thanks to the blog The Regulator, the rather supposedly unbalanced soul behind 'Bugs' has been outed. Ed Hale of Wellington, Texas, the alleged racist owner of Plains Radio, is the mysterious 'Bugs.' "[18]

Or was he?

There can be little doubt of Hale's "alleged" racism or his political extremism. Readers may recall Hale as the right-wing zealot who once called First Lady Michelle Obama a "gorilla," and who, in 2008, offered to sell audio tapes "proving" an impending presidential divorce, with bids starting at $2 million. Internet blogger "Sage" procured an alleged audiotape of Hale, in conversation with a potential buyer, and insists that Hale's voice is identical to that of "Bugs."[19] Thus far, if any scientific voice analysis has been attempted on said tapes, its results remain unpublished.

Potshots

Reports of Sasquatch shootings did not end with Bugs. Dennis Bauer reports two more widely separated cases from 1996.

> 1996 summer; AL, Macon; a dog bites a bigfoot, the bigfoot hits then dog, a farmer shoots the bigfoot JS-ABS.

> 1996-11-00; ME, Kennebec; a bigfoot falls when shot but gets up and runs away then later grunts at trailing hunter.[20]

In the first listing, "JS-ABS" refers to a short booklet, *Sasquatch: Alabama Bigfoot Sightings,* self-published by author James M. Smith in 2003, offered for sale on the Alabama Bigfoot Society's website at a cost of $13.45.[21] The second incident links back to Oregonbigfoot.com, where we read an anonymous hunter's report of a shooting near Byron, Maine, in November 1996. After hearing sounds of "something big" and spotting fifteen-inch humanoid footprints, the witness claimed the following encounter (uncorrected):

> Later that day we wanted to finish hunting so my dad waked down one road and i walked around the back side of the rock face and then when i got to the top i started to look around to see if i could see any deer or sence the deer tag came with a bear tag to maybe i'd see a bear. well when i looked down toward the bottom of the cliff and at the far end of a clearing i saw what looked like a bear about 200 yds away and i trained my 303.cal british enfield rifle on the object and aimed right about 2 inches above it's kill zone and pulled the trigger. i heard it hit and the creature fell over and got back up and stood up on two legs and ran off. i got scared because i thought i may have shot a person in a brown outfit. so i walked over to where i shot it and i looked down and seen the same tracks i'd seen earlier with my dad. so i followed them to see exactly what i shot, i'd must have followed those tracks for about half a mile then i started to notice how long the stides were and how deep the tracks went in to the dirt. (about 2-3 inches) so i knew it was about 3-4 hundred pounds easy. i knew that because i weighted 120 pounds and my tracks only sank about a quarter of a inch. i never saw any blood or hair, but there were trees that were 4 inces thick that were broken over right in line with the tracks. also i heard sounds like a low grunting like almost a warning or a challenging sounds, even though i had my rifle witch was big enough to bring down a bear, and seen what happend when i shot it, i wasn't about to see if it was just bluffing or not. i got the heck out of there and found my dad and tould him what happend and he said lets go home and find another spot.[22]

Dennis Bauer logs our only case from 1997: "1997-04-00; MS, Forrest; man playing guitar on porch shoots approaching bigfoot 4 times with a .22 targe[t] pistol."[23] The hyperlink provided leads to the Sasquatch Information Society's website with a notation "Page Not Found." An attempt to search the site produced another page advising that its database was "down for maintenance."[24]

Bauer presents two shooting reports from 1998, both occurring in November.

> 1998-11-00; TN, Robertson; hunters shoot a bigfoot that charges them at house, the bigfoot falls and crawls away.
> 1998-11-00; TX, Harrison; man shoots a bigfoot 3 times with a .22 rifle, his brother-in-law later shoots a bigfoot.[25]

A link for the first shooting takes us to the GCBRO's website, reporting an incident at Greenbrier, Tennessee, occurring at 10 a.m. on November 4. The anonymous witness was hunting deer when he found an eight-point buck "whose throat was completely ripped out and other parts of its body were also ripped off," that discovery was accompanied by "this chill like something very big was watching me" and a shriek "like no noise I have ever heard." Retreating to his truck, the hunter fled, catching a glimpse of Sasquatch in his rearview mirror. On the way home, he saw more deer in a pasture and paused to shoot one, then saw another Sasquatch—or the same one?—plodding toward a neighbor's stable.[26] He goes on to write (uncorrected):

> I realized after a few minutes that my neighbor Joel had just had two mares have foals that weren't even 2 weeks old. I didn't want it to kill them so i fired a shot in front of it hoping to scare it off. It didn't it just gazed in our direction then started running toward us. Josh hopped in his truck and told me to get in mine, but i was too scared to move, however i managed to fire off another shot hoping to scare it off, all it did was make the thing even madder. So i hopped in my truck and we took off to Joel's' house, informed him what happened then we went to my house to call the game warden and tell him about we had saw...Then i heard some shots being fired from Joel's' so i ran in and had Josh call over there but there was no answer so we got all of my guns which consisted of a 10 and a 12 gauge my 30.06 my 30.30 and of course my 308 was already out there...About then the thing yelled again and burst out of the trees not 40 feet from us. The thing was not human that we knew so we shot it. I know we hit it because it fell to the ground and started crawling away. I told Josh to stop shooting because i didn't know what it was. Later when the game warden arrived i told him what had happened and he said he had answered numerous reports lately pretty much regarding the same thing. He wasn't sure what to think about it until I showed him the blood and tracks. We followed the tracks to a creek about half a mile away from my house where they were joined by other tracks of the same kind and they went off into the creek.[27]

Presumably, none of the blood was saved or analyzed to document the incident.

The Texas case links readers to the North American Wood Ape Conservancy website, describing an encounter between two squirrel hunters and a Sasquatch. One hunter fired three shots from his .22-caliber rifle, initially saying, "I don't know if I hit it, but it let out a scream/ roar that almost made me piss myself." Later, questioned by investigator Craig Woolheater, the gunman "remembered that the subject grunted like it had been hit, then roared which went from a low roar to a high scream." The report includes no mention of a second shooting by anyone's brother-in-law.[28]

Dennis Bauer lists two incidents from 1999, as follow:

> 1999-04-00; TN, Hawkins; a farmer shoots a bigfoot with a 30.06 he kept on his tractor for that purpose.

> 1999-09-00; OK, Murray; a man shoots a bigfoot in its chest knocking it backwards, it flees screaming.[29]

Bauer's link for the first case leads readers to the GCBRO's website, for an anonymous first-person account of the shooting, reportedly occurring on April 6.

> I was plowing a field when I saw a creature walk out of the woods. It seemed to move very cautiously and look up at me where I sat on the tractor. At first, I thought it was a bear as there have been some sighted in the area in the previous months. I have kept my rifle, a 30-06, on my tractor just in case I ever ran into one of them. I took a shot at the creature and hit it. I then called my neighbor and told him what happened and asked him to come help me track it and hopefully, find its body. We searched the area until 2:00 am, finding nothing, and called off the search till daylight. In the morning we searched the property, which included several caves, but only found tracks and blood. During the search, we heard several wails that sounded like a mountain lion but much deeper and louder. We tried following the wails but never found the source from which they came.[30]

And once again, the blood was not preserved for study.

The Oklahoma case links to another GCBRO report from Dougherty, dated September 24, 1999. According to that anonymous, second-hand account:

> Friday afternoon I received a emergency email from a friend who said his son had shot a Bigfoot behind his house, while it was raiding the chicken pen. He stated he want me to come and help him find the body. When I arrived the Father of the young man who shot the creature took me out to where the incident took place and described what took place. It being too dark to track anything at that time we decided to wait until morning to try and track the creature. The next morning I interviewed the young man to get his statement on what occurred, He stated that he was leaving the house carrying a 12 gauge shotgun, while going to the bunk house to retire for the night. He stated he heard a chicken squawking out near the fighting rooster pen, he then shined a flashlight in the that direction and saw a creature holding a chicken in its hands, the young man then turned to notify his

father who was in the house and as he walked towards the house he said his hair on back of his neck stood up and he turned to see where the creature was, the creature had came towards him about half the distance when he first saw it, it had stopped in the light of a street light and was hunched down in a threatening stance, he then fired at it striking it in the upper chest area knocking it back toward the rooster pen. It then staggered back towards the road next to the property, screaming it then left the area into a patch of woods on the other side of the road. That morning I found blood at the site where it was first shot and trailed blood up on the road and down the road for a hundred yards. It was obvious the creature was badly hurt from the amount of blood on the road. The father and myself then searched the woods around the area finding no body. I have sent the samples to a laboratory for testing as of yet I have not received any results back of any testing. As I kept some of the samples I will be sending them for testing will keep all informed at a later date when testing is complete.[31]

Alas, the rest is silence. The report's confusing finale leaves us wondering how many samples of the creature's blood were saved, where they were sent, and why no test results have been announced during the fifteen years and counting since they were collected.

Oklahoma Under Siege

The Sooner State remained a hotspot for reported Sasquatch shootings at the dawn of the 21st century. Ray Crowe's IBS logs the first incident, allegedly occurring near Talihina, in LeFlore County, in January 2000. As Crowe summarized the event: "A report from Talihina tells of encounters a few weeks ago and an alleged shooting and samples of 'meat' that came from the wound and body recovery of a few days ago (fabrication?)."[32] We shall never know, since—once again—the "meat" was not preserved for scientific study.

From that dodgy report, we proceed to the so-called "Siege of Honobia," involving multiple witnesses and promiscuous gunfire in Oklahoma's McCurtain County. So many sources have described this incident—or, rather, *series* of incidents—in contradictory terms, that sorting them out takes some effort.

We begin, as usual, with Dennis Bauer, who writes: "2000-01-00; OK, McCurtain; home owner shoots a bigfoot in yard, 2 other bigfoots carry it away, FBI takes hair, blood."[33] Bauer provides a link for readers to the BFRO's website, where we find an anonymous third-hand report from Pickens, Oklahoma, dated January 28. That account reads:

I was told by phone by a relative that three large manlike, hairy creatures were observed on a person's porch and yard. The man got his SKS [rifle] and has allegedly shot one. Two others came out of the nearby forest and carried him away. That the FBI was called and samples of blood and hair have been taken for testing. My relative also told me that contractors who plant young pine seedlings are refusing to go in near Honobia and plant trees.[34]

Honobia lies on the border between LeFlore and Pushmataha Counties, fifteen miles southeast of Talihina, scene of the century's first reported Sasquatch shooting. McCurtain County is adjacent to both, occupying the farthest southeast corner of the state. Even before the Pickens incident, Honobia had Sasquatch troubles of its own, as reported by blogger Robert Lindsay.

> January 2000: Honobia, Oklahoma. The Siege of Honobia. Bigfoot apparently shot and killed as part of a group that was raiding and harassing a rural residence. Other Bigfoots apparently carried off the dead Bigfoot. Two senior and well trusted members of the BFRO were there that night shooting guns at the Bigfoots and witnessed the killing.[35]

At first glance, it appears Lindsay is simply paraphrasing the BFRO's report from Pickens, misplacing it in Honobia. As *Dragnet* narrator Jack Webb might once have said, only the name has been changed to confuse the reader. However, when we check the BFRO's site again, we find a much longer, more detailed report titled "The 'Siege' at Honobia," describing a series of events entirely separate from the Pickens shooting.

The BFRO's involvement began with an email received in mid-January 2000—i.e., before January 28—sent by an unnamed Honobia resident. That email read, in part: "Too many incidents to mention here, please have someone contact us. This is no hoax and my brother is afraid for his family. This creature is getting bolder every time it returns. This thing is huge, walks upright, smells like a musky urine, burned hair type odor. He repeatedly comes back in the early morning hours after midnight and harasses them until just before dawn. It has on more than one occasion tried to enter their home. We don't know where to turn. Everyone thinks we are crazy when we mention it. Please, we don't know what to do but I do know that something needs to be done! There are stories we could tell that would make the hair stand on your neck."[36]

On the night before that email was received, a member of the beleaguered household fired at one of the shaggy attackers, afterward finding "a substantial trail of blood in the yard." BFRO investigators rushed to the scene and questioned shooter "Tim," surmising from his story that the blood trail did not belong to a wounded Sasquatch, but to a deer carcass one of the prowlers had snatched from an outdoor shed. Still, BFRO boss Matt Moneymaker opined, "We have always known that, in the natural course of things, there would someday be an overly aggressive bigfoot that would get itself killed by someone protecting his family/ property. This may be the one, inevitably."[37]

Three BFRO members stayed at the scene, finding the carcass of a deer apparently killed by having one leg broken and its rib cage torn open, the vital organs removed. That night, unseen prowlers harassed the family and their guests with "loud vocalizations, tree thrashing, chattering and whistling outside the house," prompting several family members to fire at random from their porch. No bodies were found in the wake of that one-sided battle, though one investigator told headquarters, "The most baffling thing for all of us was why these things weren't running away after being shot at. They'd pull back a bit in the trees, then move to a different part of the hillside and could be seen through the brush when the spotlights reflected off their eyes."[38]

Wild, random shooting occurred on a subsequent evening, again in the presence of BFRO investigators, but their efforts to videotape the elusive targets proved fruitless. The group's published report makes no mention of BFRO members firing at Sasquatch, much less "witnessing a killing," as claimed by Lindsay.[39]

In search of validation for these incidents, I filed a Freedom of Information Act request with the FBI on June 30, 2014, citing the BFRO's report from Pickens and requesting any available data on federal agents collecting blood or hair samples from the scene. The bureau's reply, dated July 3, denies any knowledge of the Oklahoma incident—which it would, of course, if Washington decreed a cover-up.[40]

21st-Century Terrors

Dennis Bauer dates our next shooting case seven months after Honobia's siege: "2000-08-00; AL, Russell; camper shoots a bigfoot then abandons 'tore up' dog to the bigfoot."[41] His link delivers us to the GCBRO's website, for a report from Rood Creek Park Campground and Boat Landing, near Fort Mitchell, filed by journalist Tim Chitwood of the Columbus, Georgia, *Ledger-Enquirer*. According to Chitwood's story, a terrified young man stopped at a gas station on Alabama Highway 165 around 10:30 p.m. on an unspecified Thursday night in August, clamoring for state police. An apeman, he said, had snatched his Labrador retriever from the campground. Trying to stop it, the witness reported "firing either two rounds or emptying two clips...but the thing seemed unaffected by the gunshots." Further confusing the matter, store clerks reported that the man—who answered a game warden's questions but refused to revisit the shooting scene—drove a car with Arkansas license plates, but claimed to be from Louisiana.[42]

Bauer logs our next report from April 2001: "2001-04-00; IN, Orange; a hunter shotguns a bigfoot in face, the bigfoot is heard stumbling around in ravine."[43] Again, his link leads readers to the GCBRO's website, where an unnamed turkey hunter described his April 27 encounter with Sasquatch.

> The creature appeared to be checking the wind to locate the location of the turkey calls. I raised my shotgun just for protection due to the creature coming to within 30 paces of my location. I depressed the safety on my shotgun as the creature faced me and appeared to be stalking my scent. In an instant the creature charged and I fired one round at his face area and the creature turned and bolted down a steep raving and out of sight. I could hear the creature stumbling around in the bottom of the ravine for another 15 minutes. At this point the woods became very quiet, and I walked to the location of the shot. I observed speckling of blood scattered throughout the area where the creature exited the ravine.[44]

If any blood was preserved for testing, the report fails to mention it.

Bauer's next report, unfortunately, proved untraceable. He writes: "2001-11-00; AL, Coosa; hunter shoots a bigfoot standing under treestand, it runs away thru pasture, pond, and woods."[45] The hyperlink provided takes us back to the GCBRO's Indiana report, quoted above, with no mention of any Alabama shooting. Checking the website's Alabama reports, we find only one case from Coosa County, a July 2000 report of strange vocalizations with no Sasquatch sighted or shot.[46]

Before moving on to 2002, Bauer offers a vaguely dated incident: "2002 prior; OK, Choctaw; man shoots a bigfoot, bigfoots then throw rocks and sticks at house."[47] The link for that item takes us to the North American Wood Ape Conservancy's website, where we find a report dated October 30, submitted on December 24. It reads: "There was an old man that lived by the lake. He had shot one of them and they kept coming back and throwing rocks and sticks at his house. They found the old man dead one day, from a heart attack."[48] No further details are available.

Jumping forward to October 2002, Bauer reports: "2002-10-00; AR, Johnson; hunter shoots a bigfoot multiple times with .22 LR's."[49] The "22 LR's" refers to .22 Long Rifle cartridges, introduced in 1887, which may be fired from a variety of .22-caliber weapons. Beyond that, we have no more information on this case, since Bauer's link leads us to a suspended Web page.[50]

August of 2003 allegedly brings two reports from Texas. Dennis Bauer writes:

> 2003-08-00; TX, Walker; Police Academy student shoots a bigfoot in his backyard, the bigfoot escapes BFRO Rpt 7467
> 2003-08-00; TX, Hood; a running bigfoot knocks down a human and dog, human then shoots the bigfoot.[51]

The first report appears easy to trace, but in fact takes us nowhere, since the BFRO's website includes no report #7467. It *does* list five reports from Walker County, but none date from 2003, none bear a number resembling 7467, and none involves shooting.[52]

Bauer's second August report has more substantial roots. Linked to the North American Wood Ape Conservancy's website, it presents an incident allegedly occurring on August 18 and reported online the same day. Roused by sounds in the night, an anonymous property owner makes a "security sweep" and meets Sasquatch in the dark. He writes: "It scared the crap outta me and I got off three shots *but missed it* because my rifle jammed up with two shells in the chamber."[53] (Emphasis added.)

Two months later, from neighboring Arkansas, Bauer reports: "2003-10-00; AR, Benton; a female bigfoot gets shot by drunks at campfire in woods."[54] His link delivers us to Bigfoot Encounters, for the tale of some Fayetteville saloon bouncers, killing time in the woods of Washington County after hours. As Guy Barnes reported at NWAonline, "a female Bigfoot crashed the party," with unfortunate results. The melee ended with "the lady Bigfoot being shot and wounded and one of the bouncers being tossed across the clearing." Neither, it appears, were seriously injured.[55]

Our final incident of 2003 dates from November. Dennis Bauer writes: "2003-11-00; PA,

Westmoreland; hunter shoots a bigfoot in the shoulder, it screams and flees."[56] Bauer's hyperlink, to the Pennsylvania Bigfoot Society's website, provides a bit more detail.

> November 28, 2003, Forbes State Forest: Two men hunting from tree stands witness large upright hair covered creature walk out of the brush. One man shoots creature in shoulder. Creature runs flees and screams in pain. Two men flee shortly after creature does.[57]

No reports of Sasquatch shootings presently exist for 2004. The next year's sole event reportedly occurred in January. Dennis Bauer writes: "2005-01-00; OR, Umatilla; farmer shoots a bigfoot, the bigfoot screams, falls, gets up, and runs away."[58] Bauer's link, to Oregonbigfoot.com, dates the incident from January 14, at a farm west of Stanfield. As described by the unnamed farmer:

> I was examining my field with my nephew and I looked to my left and seen something hunched over attacking my cattle. I yelled at it to get away and it stood up and looked about 8 feet tall. I ran into my truck and got my shotgun and fired once and hit it but I don't know where I hit him. I know it wasn't a bear because when it fell and got back off it sprinted like a human. I chased its blood trail to the river where I lost it. When I shot it, it gave a creepy scream like a cry.[59]

Two shooting reports exist for 2006. As described by Dennis Bauer—

> 2006-02-00; TX, Navarro; 1) road crossing, 2) human shoots bf twice with 30.06 and it walks away *IBS*

> 2006 summer; OR, Klamath; a bigfoot gets up and walks away after being shot at with a .22 caliber rifle.[60]

While Bauer provides no link for the first case, his notation steers us to the IBS database online. Unfortunately, that trove of Texas sightings includes none from Navarro County, nor does it describe any shootings.[61]

Bauer's link for the second incident, to Oregonbigfoot.com, proves more productive, placing the incident near Chiloquin (misspelled "Chiliquin"). According to the anonymous email correspondent, her boyfriend was target shooting when—

> He came running towards me yelling my name. He is 6 feet 4 inches and he was as white as a ghost when he got to me! He said that something was up on the cliff of the hill that we have been on before. I didn't believe him until he started to shoot at it with his 22. Then after a couple of shots I saw it get up slowly turn around and start walking away. We were so scared we got in my car and took off. We came back the next day with a friend and tried to find something, but we couldn't.[62]

Patty Whacked?

Readers of *Sasquatch Down* will be familiar with the film of a purported female Bigfoot shot by Roger Patterson and Bob Gimlin at California's Bluff Creek, on October 20, 1967. They may not be aware, however, of a claim floated in early 2008 and hotly debated thereafter, alleging that the creature in question—nicknamed "Patty"—was hunted down and killed, with other members of her species, in what some writers have called "the Bluff Creek massacre."

The massacre story begins with Sasquatch researcher Marlon K. Davis, first to stabilize the rather jerky Patterson film and enlarge it using computer technology. That contribution to the field won praise for Davis from most Bigfooters, but he had more to say. Boiled down to the essentials, Davis claimed that the 53-second film viewed by millions since 1967 was only part of a longer, much more shocking reel. If viewed in its entirety, he said, the film revealed the cold-blooded slaughter of Patty and her presumed family, on par with a low-budget horror movie. Appearing at Don Keating's Ohio Bigfoot Conference on May 17, 2008, Davis opened with frames of the Patterson film, purporting to show "evidence of braids and a ponytail in the head hair," then cut to the chase with his massacre theory, receiving a chorus of boos.[63]

That harsh response drove Davis briefly "underground," with a parting comment that he had "ceased all public scrutiny of this film."[64] In November 2010 Davis emailed Guy Edwards, proprietor of the Bigfoot Lunch Club website, the following complaint: "On your comment that I have theorized that Roger Patterson and Bob Gimlin shot bigfoot at Bluff Creek. This is absolutely untrue. I don't know where you got that from but I have formulated no such theory. I respectfully request that you correct this assertion that I have accused Patterson and Gimlin of this, as I have not done so."[65]

Between those two statements, in July 2009, a Davis associate named David Paulides told Sasquatch hunter Al Hodgson, "I actually got my hands on a fairly old copy of the PG film, full framed with segments on it nobody has seen. It is in the experts hands and many of our impressions of what actually occurred is playing out." Paulides also claimed that Canadian researcher John Green harbored a "very, very dark secret" concerning the Patterson film.[66] A month later, writing directly to Green, asking a series of accusatory questions that included:

> What is the man pointing to on the ground that you appear to be looking down at with something on your shoulder?

> What is on your shoulder?

> John—you say that isn't a camera on your shoulder? What is it then?

> We had our experts determine what was on your shoulder and with 99% accuracy they determined it was the attached and inserted camera. I trust your integrity, but please tell me what is on your shoulder?

> It appears you have something on your belt just above and in front of your left elbow. What is it on your belt? Is that a pistol holster?[67]

In fact, a viewing of the entire film in question demonstrates, among other things, that Green has *nothing* on his belt, while the object "on his shoulder" is in fact a small hand-held camera. More significantly what Paulides and his unnamed "experts" failed to realize is that the film they used to pillory John Green and others had been shot at Blue Creek Mountain, in Del Norte County, where more than 1,000 Sasquatch tracks were found in August 1967, two months *before* the Patterson sighting. Green did not visit the Bluff Creek site until June 1968, eight months *after* Patterson's filming. Needless to say, Green's Blue Creek Mountain film, viewed in its entirety, reveals no Sasquatch corpses, "bloody hands," "bloody dog prints," or any of the other so-called "evidence" touted by massacre fabricators.[68]

Sasquatch conspiracy theorist Doug Tarrant added a new twist to the massacre fable in December 2011, writing to Steven Monk of the Georgia Bigfoot Society. Purportedly answering some unnamed correspondent's claim that the "Minnesota Iceman" was one of the creatures shot at Bluff Creek, Tarrant wrote, "The Iceman was shot a year earlier in Minnesota. I have personal friends in Winona Minnesota who knew Frank Hansen who shot the creature in 1966, and stored the body in a block of ice that he kept in his barn on his farm."[69]

Friends or not, we've seen that according to Hansen, writing for *Saga* in 1970, he allegedly shot his Sasquatch in the winter of 1960, nearly seven years before Patty's film debut. From that shaky beginning, Tarrant wrote to Monk (unedited):

> FACT: It was several years later and much more experimenting with that same film that M. K. Davis found what was rumored as PROOF that Pattie was shot in the leg and that a possible "massacre" had taken place with bodies left back at the site where Pattie was walking away from. This caused a lot of problems and people like John Green and others disputed that it ever happened. Later there was a big "cover up" that came about.
>
> FACT: M. K. called me and said he was going to be in the Lake tahoe area for skiing and wanted to show me his "proof" on his laptop computer. The day we were to meet, his buddy hurt his leg in a skiing accident and M. K. had to drop everything and rush him to the hospital and we weren't able to get together.
> FACT: Loren Coleman posted on his blog site that no such massacre happened and that M. K. Davis was full of sh!t!! M. K. attended a BF conference and showed a slide show of his evidence. It was too hard for everyone to believe and M. K. was boo'd. Loren then had John Green and others say that it wasn't so (?) M. K. then dropped off the BF sites and went underground. Only a few knew where he was. M. K. then got a photographer and they proceeded to make a movie of the evidence. years passed
>
> Rumor had it that the film was covered up by the government. (?)
>
> Keep in mind the government prefers to keep Bigfoot a MYTH like they do UFO's.[71]

None of which proves anything except, if Tarrant is correct, that Davis ducked one opportunity to show his "proof." The Davis film, if it exists, remained untraceable at press time for *Sasquatch Down,* As for the rest, we may as well quote Shakespeare from *Macbeth,* Act 5, Scene 5: "It is a tale told by an idiot, full of sound and fury, signifying nothing."

Dyer Straits

The 21st century's most notorious Sasquatch hoax dates from June 2008, when rumors of a Sasquatch kill began to circulate in northern Georgia. The men behind the story, used car salesman Rick Dyer and Clayton County policeman Matthew Whitton, created a website on June 16, using the domain name "bigfoottracker.com." Within days, videos began to surface on YouTube, posted by "RDYER678," claiming that a friend of Dyer and Whitton—an unnamed "felon"—had recently slain a Sasquatch, using a .30-06 rifle.[72]

Real-life drama intervened on July 3, when Whitton suffered a minor gunshot wound while grappling with a robbery suspect on duty. Whitton received local media coverage and left the hospital on July 9, just in time to post a YouTube video claiming he and Dyer had recovered the Sasquatch cadaver. On July 23 they held a press conference, hyping discovery of the supposed primate corpse.[73]

Enter researcher Steve Kulls, employed as recently as March 2008 by Sasquatch hunter—some say "hoaxer"—Carmine "Tom" Biscardi. Before proceeding with the case at hand, a few words are in order on Biscardi's background. According to his official biography, Biscardi became obsessed with finding Sasquatch after viewing the Patterson film on television, in December 1967. By 1970 he was a protégé of Sasquatch film hoaxer Ivan Marx, establishing Amazing Horizons Inc. the following year to distribute Marx's footage. In December 1973, at age 25, Biscardi got his first national exposure from *Saga* magazine, in an article touting his plan to capture a Sasquatch in the near future. Speaking from his "luxurious San Jose, Calif., apartment," Biscardi awaited "the latest word on any sightings from Ivan Marx and the other six members of the Alaskan expedition."[74]

That word never came, but Biscardi persevered. His former predictions were long forgotten by May 1981, when Biscardi told the *San Francisco Chronicle* of his personal Sasquatch sighting on nearby Mount Lassen. Again, Tom forecast an impending capture that never occurred. Two years later, as head of Amazing Horizons, he produced and distributed *In the Shadow of Bigfoot,* chockfull of phony Marx footage and marketed as "good for entertainment value only." Biscardi then dropped off the Sasquatch map until January 2004, when he resurfaced seeking a million-dollar bankroll from corporate sponsors, to pursue an albino Sasquatch allegedly photographed by "tracker Peggy Marx" (Ivan's widow). The rest was silence until April 2005, when Biscardi and Marx's grandchildren claimed to have seen a Sasquatch near Burney, California. Alas, on that occasion, Biscardi had "left his tranquilizer gun and his wire-mesh grenade launcher at home."[75]

Undaunted by that failure, in June 2005 Biscardi produced a press release headlined "Imminent Capture Anticipated." Home viewers could watch the action online, at Biscardi's website, by paying $19.95 per week or $59.95 for the expedition's full run. His plan, Biscardi said, was to incapacitate Sasquatch with a tranquilizer gun, hold it for a 90-day medical examination, then release it to the wild unharmed. Subscribers witnessed none of that, though Biscardi appeared on *Coast to Coast AM,* on August 18, telling host George Noory that his (Biscardi's) team had captured a teenage male specimen, offering live video clips at new rates. Biscardi himself had not seen the creature, and the lie collapsed on August 23, with Biscardi saying he was "hoodwinked" by a colleague of Peggy Marx. In June 2006, Biscardi claimed to have a "Bigfoot hand," found by Montana police, but X-rays proved it was a black bear's paw. One month later, Biscardi sued the Great American Bigfoot Research Organization—founded by himself and Peggy Marx—claiming its officers had promised him $250,000 to help them find Sasquatch, then shorted him $185,000 and stole his hunting equipment.[77]

Meanwhile, Rick Dyer faced some difficulties of his own. Born in 1967, he found a job selling cars in Atlanta, then afoul of the law in 2001, when an auto finance company won a $15,000 default judgment against him. Dyer joined the army in 2002 and served two years, emerging in 2004 to become a state prison guard, while his ex-wife sued for child support. Dyer lost that job in 2006 and returned to selling cars, but his losing streak continued. That same year, one of his customers won a $3,800 default judgment against Dyer for selling him a "broken" Chevrolet Corvette.[78]

Flash forward once again to July 25, 2008, when Steve Kulls invited Dyer and Whitton to appear on his Internet radio program. Dyer turned up solo, three days later, describing his Sasquatch carcass as eight feet eight inches tall, nonetheless claiming that if it was shaved and wearing a hat, it would be indistinguishable from any man on the street. On July 29, Dyer told Kulls that he had sold the creature to the *National Enquirer,* but he retracted that lie one day later, appearing with Whitton on Tom Biscardi's separate Internet show. Biscardi flew to Georgia on August 1, and on August 2 told Kulls he was "right there when the DNA was cut from the body," adding that "this thing has incredibly thick skin." Later, Biscardi revised that account, saying that Dyer and Whitton met him in a hotel lobby, handing him the sample.[79]

From there, it was off to the races. While Kulls declared himself bound by a "gag order" forbidding further comment, the Georgia tissue sample found its way to Dr. Curt Nelson, a biology professor at the University of Minnesota. Biscardi described the Sasquatch corpse to Kulls, dwelling on its teeth, eyes, and genitals, while saying that two unnamed Russians were flying in to perform a necropsy—film of that procedure to be auctioned off, with bids beginning at $11 million. On August 11, the same day black-and-white photos of the carcass "leaked" to YouTube, Dr. Nelson returned a preliminary DNA verdict, saying the sample was consistent with an ape or human. Color photos of the "creature" surfaced on August 12, followed immediately by cries of fraud. Biscardi countered on the 14th, insisting, "It's not a costume. It's the real thing."[80]

The balloon sprang more leaks on August 15, at a California press conference where Biscardi joined Dyer and Whitton, summarizing a confused DNA report and verbally assailing critics who suspected him of yet another hoax. Whitton offered a third, contradictory description of the beast's discovery, while Biscardi claimed to have touched and smelled the carcass. Before day's end, Dyer and Whitton appeared on MSNBC, Dyer touting his plan to sell the "body" to the highest bidder. Finally, on August 16, Indiana researchers thawed the pseudo-Sasquatch from its tomb of ice, revealing the predicted monster costume with synthetic hair and rubber feet, stuffed with meat and offal from a butcher's shop.[81]

On August 18, 2008, Dyer and Whittton returned to MSNBC, admitting their parts in the hoax. Whitton's police chief fired him the next day, on charges of fraud. By August 20, both hoaxers were blaming Biscardi for "coaching" their lies, blaming him for the whole tangled web of deception. Kulls appeared on *Fox and Friends,* blasting "the deceptions of Tom Biscardi," while the creature's rumored buyer—William Lett Jr. of Indiana—threatened fraud charges against all concerned, seeking return of his $50,000 purchase price.[82]

Dyer, unconcerned, told CNN on August 21, "It's just a big hoax, a big joke. It's Bigfoot. Bigfoot doesn't exist." Whitton agreed, saying, "It started off as some YouTube videos and a website. We're all about having fun."[83] Perhaps it was another bit of "fun," in January 2011, when Dyer found himself jailed in San Antonio, for swindling various eBay shoppers with items he sold but never delivered. He posted $10,000 bail, but no information is available on the legal disposition of that case.[84]

Two years later, in January 2014, Dyer returned with—you guessed it—yet another Sasquatch "corpse." Although he once declared the creature nonexistent and a "joke," Dyer now claimed to have shot a specimen himself, near San Antonio, in September 2012. He was prepared, at last, to tour with the corpse, while hawking video copies of a documentary film titled *Shooting Bigfoot.* Putting on an earnest face for journalists, Dyer declared, "Bigfoot is not a tooth fairy. Bigfoot is real. The most important thing to me is being vindicated, letting people know that I am the best Bigfoot tracker in the world and it's not just me saying it. Every test that you can possibly imagine was performed on this body, from DNA tests to 3D optical scans to body scans. It is the real deal. It's Bigfoot and Bigfoot's here, and I shot it and now I'm proving it to the world."[85]

Well, not quite.

This time, the "corpse" that Dyer placed on tour—dubbed "Hank" for reasons still unclear—was even less convincing than the frozen suit in Georgia. By February 2014, after raking in some $60,000 from sideshow patrons, Dyer confessed to yet another hoax. "I never treated anyone bad, I'm a joker, I play around, that's just me," he averred. "From this moment on, I will speak the truth! No more lies, tall tales or wild goose chases to mess with the haters!" In Spokane, Washington, Chris Russell—proprietor of the Twisted Toy Box custom costume shop—admitted crafting "Hank" for Dyer, out of foam, latex and camel hair.[86]

Through it all, Dyer seemed supremely pleased with himself. "It's really easy to trick people," he said. "There's no more evidence for Bigfoot than the Tooth Fairy or the Easter Bunny. And that's what people have to get through their heads."[87]

The Sierra Kills
Between Rick Dyer's first and second Sasquatch hoaxes, another double slaying allegedly occurred on October 8, 2010, near Gold Lake, in California's Butte County. Confessed triggerman Justin Smeja was bear hunting with a friend—known only as "the driver" to this day—when they spied an adult Sasquatch through binoculars, standing at roadside. Smeja shot it with his rifle, watched it fall, and then advanced to the site where two "babies" scrambled around their dead mother on all fours. Smeja then shot one of them, killing it instantly,

whereupon the second "baby" fled into the woods and disappeared. Smeja and his unnamed friend did likewise, racing to their pickup truck and off to reach the nearest town. Before they left, however, Smeja and his pal "cut out a piece of steak" from the adult specimen, for later scientific testing.[88]

A full year passed before Smeja went public with his story, interviewed online by Sasquatch blogger Shawn Evidence. Sasquatch researcher Derek Randles—billed by Evidence as "Justin Smeja's boss"—supported the story, insisting that "this is no hoax." Randles claimed to have a DNA report on the Sasquatch "steak," but could say no more under terms of a legal nondisclosure agreement. "It's so hard for people to believe that we are serious researchers with good true intentions," Randles acknowledged. "We are! This isn't Georgia, that I will promise you, and we sure as hell aren't the Georgia boys. You'll see."[89] Smeja's hunting partner, questioned separately, said, "I do have my part of what we recovered. I'm holding onto it."[90]

Lacking any immediate physical evidence, Smeja and Randles made do with a description of the creatures, reconstructed from memory. Smeja often seemed confused, as when he claimed the adult specimen "was so much like [Roger Patterson's "Patty"] but totally, totally different and not even close in appearance." In fact, it seemed "as different from Patty as Chuck Norris and the cable guy." It weighed about six hundred pounds, compared to forty for the young ones, which had "rounded" mouths and "talked" like humans, "in the vocal way of the deaf, with no ape like grunts." Facially, the young creatures were "somewhere in the middle between human, ape and boxer [dog]."[91]

Smeja's Sasquatch "steak" eventually found its way to Trent University—a liberal arts and science-oriented institution located along the Otonabee River in Peterborough, Ontario, Canada—and to DNA Solutions Inc., a private laboratory in Oklahoma City. According to "Sierra Evidence Initiative" spokesman Bart Curtino, each lab received a frozen sample of Smeja's kill, plus "a piece of the 'salt-cured' tissue originating from the same mass," furnished "for purposes of checks and balances."[92] At Trent, lab supervisor Dr. Bradley White reported that—

> It is considered that the species of origin that is the major contributor of nuclear and mitochondrial DNA in [this] sample...is a female black bear, *Ursus americanus*....The sample also yielded human mitochondrial DNA with a control region sequence.... Analysis of the human mitochondrial control region sequence...indicates that it is a European haplotype. It appears most frequently in East Europe and Caucasus...[where] it initially originated in the Caucasus mountains region between the Black and Caspian Seas.[93]

For the record, genealogical research reveals that members of the Smeja family immigrated to the U.S. in the 19th century, from the Austro-Hungarian Empire, with dual capitals in Vienna, Austria, and Budapest, Hungary.[94] At its peak, the empire sprawled over some 261,242 square miles, and while it never reached the Caucasus, some of its citizens were immigrants from farther east.

The report from DNA Solutions, signed by Director Brandt Cassidy, confirmed the Trent findings. To wit,

> The frozen tissue sample was tested first. DNA was isolated and PCR analysis for the hyper variable regions of the mitochondrial genome and the Cytochrome B gene region. The isolated DNA and PCR reaction products were evaluated by gel electrophoresis below. These PCR products were sequenced and determined to be of human origin. The ancestry of this mitochondrial sequence indicates a T2 haplotype. The DNA was also tested for human nuclear DNA. This sample produced a partial profile consistent with a human male. This profile was then compared to a reference sample for Justin Smeja and was determined to be identical. This is consistent with the majority of the amplifiable DNA isolated from the frozen tissue sample having originated from the same source, Justin Smeja.

> The second sample, dried tissue, was also tested through the same analysis. These PCR product was [sic] sequenced and determined to be of Ursus americanus (Black Bear) origin. The DNA was also tested for human nuclear DNA. This sample produced a partial profile consistent with a human male. This profile was then compared to a reference sample for Justin Smeja and was determined to be identical. This is consistent with the majority of the amplifiable nuclear DNA isolated from the frozen tissue sample having originated from the same source, Justin Smeja.[95]

In short, no Sasquatch and no reason to believe Smeja's story. September 2013 brought reports that Smeja had faked another Sasquatch sighting, hiring a Hollywood makeup expert to create a monster for a forthcoming feature film.[96] Despite revelation of that hoax, the 65-minute film premiered in California two months later, titled *Dead Bigfoot: A True Story*.[97] Depicting the alleged "Sierra kills" of October 2010, it may be viewed online today at Smeja's website, at a cost of $3.99 per screening.[98]

Dates Unknown

Finally, the record of Sasquatch shootings reported online includes 18 incidents described without dates, and in some cases without locations. As we shall see, however, some of those are duplicated from reports already covered in preceding pages, listed redundantly through careless website editing.

Dennis Bauer reports an incident from the Pelican State as follows: "Unknown; LA, Desoto; dogs and a bigfoot fight, a hunter shoots the bigfoot IBS."[99] His source provides more second-hand details, from researcher J. Frank McAneny: "A Desoto Parish school bus driver told me his father came face-to-face with the monster while coon hunting one night. The creature attacked the hunter and his dogs. The hunter fired point-blank at the monster but failed to stop its onslaught. The hunter's dogs still refuse to go into those woods under any circumstances."[100] One may well ask how the hunter and his dogs survived, after he "failed to stop its onslaught," but no answer is available.

Bauer's next report reads: "Unknown; OH, Adams; a bigfoot throws rocks at human, a human shoots the bigfoot, blood is found."[101] A link directs readers to Oregonbigfoot.com, quoting an unnamed resident of West Union.

> I was checking the creek to see how much water there was in it, and something threw a rock at me. So I fired my shotgun at it. I heard it run away and I couldn't catch up to it, although it left a trail of blood behind. The tracks were humanoid it shape but the size was greatly exaggerated. I went home to tell the neighbors, and when I came back I couldn't find the footprints anymore, but I did find a little blood on a leaf.[102]

Our first undated duplicate report from Dennis Bauer reads: "Unknown; OK, Le Flore; bigfoots pound on house, raid freezer, human shoots a bigfoot IBS."[103] It requires but a glance at the IBS database of Oklahoma sightings to recognize the incident covered earlier in this chapter, from January 2000.[104]

Likewise, Bauer's next case is another retread: "Unknown; WA, Clark; man shoots a bigfoot and tries to sell it until informed of the Washington no-kill law IBS."[105] This is, of course, the hoax put forward by habitual liar Ray Wallace, which opened Chapter 4 of *Sasquatch Down*.[106]

Bauer's next account initially appears to stand on somewhat firmer ground: "Unknown; WV, Harrison; coon hunter shoots and kills possible bigfoot."[107] His link takes us to a second-hand story, shared by an anonymous hunter on Bigfoot Encounters. The alleged Sasquatch killing followed a month of nightly visits that disturbed a Bridgeport, West Virginia, farmer and his hunting dogs. At last, the report (uncorrected) declares:

> One night the dogs treed an animal and the farmer got there quickly and shot it. To his surprise it was no coon. It had long grayish, brown hair and was about five feet tall. It's hands were human like and it's feet was more hand like then anything. I told my friend he was crazy so he decided to prove it to me. He told me the old man kept the animal but did not have it mounted cause he was afraid he had done something wrong. My friend took me to the old barn and there it was. The old man had nailed it's carcass to the wall. I was shocked it was built a lot like a human had hair 6 or 7 inches long on it. It had very large sharp teeth and resembled some kind of monkey looking creature. I told my friend I wouldn't say anything about it but I feel that it is my duty to report this. If anyone has any idea what this animal could be please let me know.[108]

While summarizing that report, Bauer seemingly missed the disclaimer appended by Bobbie Short, reading: "Jack Blevin of the West VA paranormal investigations believes this story is untrue as informant refused a request for photographs or onsight [*sic*] analysis of remains. Further investigation revealed email address not good and this 'story' had been around the block a few times on several different websites...finally written off as an urban legend."[109]

Blogger Robert Lindsay offers our next case, writing: "Unknown date: Klakas Inlet, Southern Alaska. In far southern Alaska on Prince of Wales Island, a Bigfoot was reportedly shot and

buried at the mouth of a stream on the north side of the inlet. Reported in the Bigfoot Track Record [*sic*]."[110] Ray Crowe did, indeed, report this incident—in fact, he found it so impressive that his IBS database logged it three times, under three separate case numbers.
Crowe credits the tale to Peter Byrne in two reports, and to researcher Rob Alley in the third. All three accounts agree that the creature was shot at Klakas Inlet, on Prince of Wales Island, part of the Alexander Archipelago in the Alaska Panhandle. The third and longest version claims the incident occurred sometime around the turn of the century, presumably in the 1890s or early 1900s.[111] According to that summary:

> One native on Prince of Wales Island...happened to exit the forest above a beach to be confronted with the sight of one of these creatures standing at point blank range. The man is supposed to have fired away, killing the creature. Three other men of his group responded to the noise and, the tale goes, found the creature lying there "so horrible that they buried him on the beach." What is not said is whether the burial took place out of fear of reprisal by the "land otter people," out of remorse for killing a creature so superficially manlike, out of fear of repercussions by the non-natives, or even the non-native authorities.[112]

Robert Lindsay's next report is the epitome of vagueness: "Date unknown, modern era: Location unknown. A wealthy hunter reportedly shot and killed a Bigfoot, then paid a taxidermist to stuff it, and it's presently on display in some ritzy country club on the East Coast. Reported by Ray Crowe."[113] Crowe, blogging in March 2011, wrote: "Ron [?] reports that there is a report that some rich fellow shot a Bigfoot, sent the body to a taxidermist, and the Bigfoot is in some kind of east coast country club (skepticals on please)."[114]

Lindsay follows with a case containing slightly more detail: "Unknown date, modern era: Yankton, Oregon. Near the Colombia River north of Portland, a hunter shot a Bigfoot four times between the eyes and killed it. It rolled off the road. The man came back 24 hours later, and the body was gone. There was a set of three tracks, possibly a family group—a male, a female and a juvenile. Reported by Ray Crowe."[115] In fact, Crowe apparently mentioned this shooting for the first time in March 2011, in a blog posted only two months before Lindsay quoted it nearly verbatim. Crowe's article credits the story to "Jim," otherwise unidentified, speculating that the dead creature's mate absconded with the corpse.[116]

In that same blog, Crowe sketched another undated case, credited to late researcher Datus Perry, who claimed more personal Sasquatch sightings than any other living person.[117] According to Crowe, "Perry said a hunter near Green River, Washington, saw a bear digging for grubs. He shot it, and saw to his horror he had killed a 'hairy-man.' He rolled a log over the body and did not report it until on his deathbed years later."[118]

Shifting to the Southwest, Crowe wrote, "Northwest of Gila Dam, Arizona, the witness shot a creature. The female Bigfoot had been standing right above him while he slept. When he woke and grabbed his rifle, it fled, but he shot it anyway."[119] Apparently, the shooter missed, or at the very least did not inflict a mortal wound.

Another case devoid of date or location also comes from Ray Crowe. He wrote, "There was a nine foot Bigfoot wading in six feet of water. When it charged the witness, he fired a shot at point blank range from ten or fifteen feet away. It let out a god-awful scream when it was hit in the chest, and the witness scrammed. Later a game warden tried to arrest him for shooting an endangered animal. He replied by replying, there was '...no such animal, so how could he shoot an endangered species?' "[120]

With his penchant for first-name dropping, Crowe credits his next report simply to "Dan." As detailed in his blog of March 2011: "Dan was on the Warm Springs Indian Reservation, Oregon, as a boy. He heard a farmer tell of shooting a Bigfoot that he brought home with him. That evening several creatures visited and threw boulders at the house and buried it (like at Ape Canyon). The farmer got away, but the Bigfoot body was taken. The buried house is still visible near the motorcycle track in the town of Warm Springs."[121]

Robert Lindsay's next report reads: "Unknown date, modern era, Amboy, Washington. Near Mt. St. Helens, a hunter reported that he shot and killed a male Bigfoot on an old logging road. Upon hearing that there was a $10,000 fine for killing a Bigfoot, the hunter hung up the phone on the researcher. Reported by Ray Crowe."[122] In fact, Crowe says he took the unnamed shooter's call himself, and "told him that by law there was a ten thousand dollar fine in Skamania County for killing a Bigfoot. The caller hung up. Chuckle—Amboy is actually in Clark County."[123]

Robert Lindsay dips south of the border for his next item, writing: "Unknown date: Sonora, Mexico. Rich Grumley reported that a hunter shot and killed a Bigfoot, then buried it."[124] Sadly, as noted in Chapter 2, California researcher Louis "Rich" Grumley died in 2000 and was thus unavailable to comment on cases reported in his name by others.

Ray Crowe provides the next item of interest, worth quoting here in its entirety.

> From a Bob Slaughter book. T-Jean reports a 400# Bigfoot shot by hunters with a dog pack and buried in an old well 20 miles south of New Orleans near Lafitte. It was an oversize hairy human and they feared they might be prosecuted. The seven foot skeleton with huge canine teeth in a brow-ridged skull was supposedly recovered later by Bob and students of Tulane University where the skeleton was taken afterwards, then sent to Ladonia University. Does not exist—a cover for secrecy? There is a Ladonia in Texas. The skeleton had a quartz crystal on a chain (sure, sure). Some of the students still live in Lafitte and supposedly have photos. I do not really know what to make of this report. You judge.[125]

From that beginning, I was able to improve somewhat on Crowe's research. While Crowe concluded that Ladonia University "does not exist," Texas journalist Rob Davis found it easily enough in June 1993, located in the aforementioned town of Ladonia, 80 miles northeast of Dallas. Like its town (population 700), LU was a tiny institution, founded by retired Southern Methodist University paleontology professor Bob Slaughter and his wife in 1988. Aside from teaching and collecting Fortean relics, Slaughter was also writing a book, titled *Fossil Remains of Mythical Creatures,* published in 1996.[126] Davis describes the pertinent matter:

Over fava beans and hummus at Ali Baba, one of Slaughter's favorite cafes, he's telling me about Rugaru-the half-man/half-wolf of Louisiana, and I don't mean Harry Connick Jr. This hairy, big-toothed beast was 8 feet tall and weighed about 500 pounds and was brought down in 1921 by Cajun trappers and their dogs in a bayou in Lafourche Parish. (Slaughter has a map; he has maps of every fossil site.) Scared that they might get in trouble with the law, the trappers had shoved the huge, half-human body into a well and left it. As luck would have it, no small child from West Texas named Jessica ever did a header into the well, so the ad hoc grave and its gruesome contents remained a swamp secret. Slaughter only learned about it through T-Jean Ledet, a graduate student who had come to hear the professor lecture at Tulane in the early 1970s. Eventually T-Jean led Slaughter to the well, where they hoisted the skeletal remains, plastered them and shipped them back to Ladonia. "The odd thing," Slaughter recalls, "is that we found a quartz crystal tied to a bottle chain around its neck. Now, the only place that crystal comes from is in the Ozarks-which means he'd been up there and probably accounts for the Big Foot legends of that region.[127]

While viewing other specimens in Slaughter's personal cabinet of curiosities, Davis did not see the massive skeleton in question. The question of its actual existence thus remains unsettled.

Ray Crowe next takes us to the Sooner State, with this report: "The Bigfoot was named the 'Abominable Chicken Man,' because it kept killing and eating local chickens. Tim, from Kiamichis [sic], Oklahoma, saw four of them come out of the woods, and he shot one of them. There was no body, and Tim said that the Bigfoot carry off their dead and bury them."[128]

Finally, we have one last report from Clark County, Washington, logged by Ray Crowe, credited to researcher Larry Lund. The sketchy item reads "BF SHOT, BLOOD TRAIL AS RAN OFF," but the present IBS database online does not include the promised "full report."[129] Lund, like the aforementioned Rich Grumley, died in 2000, and the trail ends at his grave.

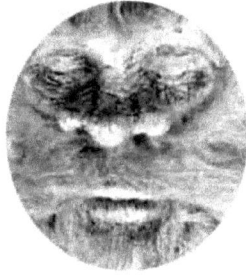

Chapter 6.
Road Kills

R andy Lee Tenley, age 44, felt like playing a prank on August 26, 2012. Donning a "ghillie" suit, normally used to conceal military snipers from their enemies, he stood beside U.S. Highway 93 that night, south of Kalispell, Montana, primed to frighten passing motorists. As the first car approached, Tenley leaped onto the pavement—and was flattened by the vehicle. A second car, coming along behind the first, ensured that he was dead. State Trooper Jim Schneider interviewed Tenley's friends, telling reporters, "He was trying to make people think he was Sasquatch so people would call in a Sasquatch sighting. You can't make it up. I haven't seen or heard of anything like this before. Obviously, his suit made it difficult for people to see him." Schneider also allowed that "alcohol may have been a factor" in Tenley's unintended suicide.

Comic, tragic, or a bit of both? In any case, Tenley's fate begs the question: if Sasquatch exists, why has none ever been found as roadkill?

America's highways are open-air slaughterhouses. Some 40 million animals of all sizes and species are run down each year, an estimated 20 *billion* individual kills between 1952 and 2012, although precise tabulation is impossible. Five billion of those were large mammals such as bears, coyotes, deer, elk, moose and wolves, accounting for 26 percent of the total death toll in any given year—approximately eight roadkills per hour across North America. In 2006-07, 26 Florida panthers from a population of 100 died on state roads. Florida also loses an average of 150 black bears each year, from a known population of 1,500. Ohio lost 25,636 of its half-million deer to motorists during 1993-94, and so on.[2]

Where, then, skeptics ask, are the Sasquatch bodies scarred with tire tracks? If hairy manimals roam every mainland U.S. state and Canadian province, as indicated by sighting reports, how do they avoid joining every other beast that walks, crawls or flies on the daily casualty lists? If none are found, is that not proof in fact that none exist? A blogger calling himself "The Other John McC," self-described as "a full-time research scientist for a major US government laboratory, with a Ph.D. (ABD) in experimental psychology research," estimates that if thousands of Sasquatches roam the continent, 32 to 96 should be run down each year—a range of 1,920 to 5,760 road-killed since the early 1950s.[3]

So, where are they?

Sasquatch believers offer various replies to that challenge. Reference to their rarity will not stand, since some elusive species are seen only after being killed by accident. Likewise, suggestions that manimals avoid highways are clearly specious: at least 178 highway sightings were logged between 1927 and 1980, in 38 states and five Canadian provinces, with that number increasing yearly.[4] Is Sasquatch "too smart" to be hit? It seems doubtful, when 110,994 human pedestrians died in collisions between 1990 and 2011, with many times that number injured.[5] The BFRO opines that "Bigfoots are extremely rare and extremely cautious—so much so that the odds of a roadkill have not caught up with any yet."[6]

In fact, however—if reports can be believed—bipedal cryptids *have* been struck and either killed or injured on no less than 22 occasions between the late 19th century and 2001.

Sasquatch vs. the Iron Horse
Before automobiles existed, one Sasquatch reportedly met its untimely end beneath the crushing wheels of a train. Robert Lindsay writes: "1880's: Montana, near the Canadian border. A Bigfoot was killed by a train, and its body was stuck under the train. Reported by Rita Swift."[7] Thankfully, Swift's tale is preserved online, after first appearing in Ray Crowe's *Track Record.*

> My name is Rita Swift. I live in Orange Co. California. In 1945, my grandfather George Huhn told me a story about the time his train hit a large Ape creature and bent the cowcatcher on his train. This was in the 1880's and he was an engineer on a train that ran along the borders of the US and Canada. It was night, and all of a sudden their train hit something and they stopped the train, because the cowcatcher was dragging on the tracks.
>
> At first they thought it was a moose, but when they all got out with their lanterns, they discovered this huge smelly Ape, hung up in the catcher. They had only lanterns for light, and they were in the forest, basically in the middle of nowhere. It took most of the crew to pick it up and lift it into an open flat car. They noticed it was structured differently from a Gorilla or Ape, and smelled so bad, the crew got the smell on them. They left it on the flat car, because it took at least 2 hours to straighten out the cowcatcher. Good thing my great grandfather was also a blacksmith. They were at least 2 hours from the next water tower and station of sorts. The brakeman noticed Indians sneaking around in the forest, but thought they had disappeared. When they were ready to go, the crew checked on their smelly passenger, but he was gone. They looked for tracks and decided the Indians had dragged it away into the forest and across a stream. They found the tracks and

pieces of hair and of course the smell. They washed up in the stream and were glad to get rid of it. The smell had even remained in the flat car. My great grandfather took pieces of the hair back, and gave it to a doctor he knew in Michigan. They had all decided the creature had escaped from a circus or sideshow. Great grandfather thought it was 8 feet tall and weighed at least 500 lbs. It took six men to carry it off the tracks. When my daughter was a student at California State University at Fullerton in 1986, I met a Professor of Anthropology...I noticed in her office she had information on the walls about Big Foot. I told her the story and she believed it was documented.

My grandfather said the Ape had a different face than what he remembered of a Gorilla. He said the teeth were like humans [sic] , but extremely wide and large. The body hair was thick dark brown, with light tipping and the eyes were large and dark. He said they agreed it was a male because of it's genitalia...

Grandfather would never tell stories that were not true. He was a devout Methodist, and said his prayers so loud every night, the whole house could hear him. He had originally come from Amish in Mercer, Co., Pa., but left to fight for the Union in the Civil War. His father did not accept his decision, and he never returned to Mercer Co. He was born in 1845 and died in 1947, in Claremont California.[8]

Eight decades passed before our next report of a train striking Sasquatch. Robert Lindsay writes: "January, 1965: Chemult, Oregon. On the Umpqua National Forest northwest of Crater Lake National Park, a Southern Pacific train traveling between Bend and Klamath Falls towns hit a Bigfoot and killed it. The crew never reported it to their company or authorities because they thought they would be accused of drinking on the job."[9] Ray Crowe considered that "a fun report" and logged it into his IBS database.[10] There, we read:

"The Great Freight Train Caper," Report from Michael Jay. Got this one from my brother-in-law, who works for the railroad and he heard it from the involved engineer. Bigfoot was clobbered and knocked into a canyon years ago by a southbound Southern Pacific freight train highballing through the Cascades. Never did get the engineer's name, but the incident occurred in January of 1965, if memory serves. The location was the SP's southbound main line somewhere near Chemult, Oregon, and the train was en route to Klamath Falls, the crew-change point for California-bound trains. The time was early evening, just before dusk, and the train was said to have just topped a grade and started picking up speed when the engineer observed "a large man in a fur coat" standing on the tracks ahead. It was snowing, so I'd assume that visibility was less than ideal...must've been a fairly close-range observation. The engineer had time to do little else besides blow his whistle before impact, and the "Man" was sent flying into a steep canyon which abutted the railroad grade. Naturally the train was stopped and a brakeman or two went down to find the body...which turned out, of course, not to be "human" at all, but rather an enormous, shaggy, ape-like being which appeared stone dead. The incident was allegedly never reported for fear of disbelief and/or the crew's being accused of drinking on the job. My brother-in-law heard this story in the early 70's from the involved engineer, who happened to be passing through Eugene, but was not stationed there...hence part of the difficulty in obtaining a more complete report. My brother-in-law often comes into contact with the train crews (he's Foreman for SP's Mechanical Dept. car shop), though he doesn't know most of 'em personally aside from a handful whose runs happen to originate in Eugene. This guy was probably from Albany or Portland. As to the validity of the story, my brother-in-law said at the time that the gentleman seemed quite serious and matter-of-fact about the occurrence, and he noted that trains hitting deer and elk during their trip along the Cascades line is a fairly commonplace event...so why not a Sasquatch? Makes sense to me.[11]

Finally, concerning trains, we have an undated report from the late Rich Grumley, summarized by Ray Crowe as follows: "Rich said an Amtrak train was at the depot in Omaha, Nebraska, and the front coupler of the lead locomotive was covered with a dark-colored mass of flesh. He thought that it might have been a Bigfoot hit at 80 mph, but the train left before he could collect samples. The crews tell of hitting some kind of animal almost every trip."[12] Hardly conclusive evidence, but offered here for what it may be worth.

Highways to Hell

There are more cars and trucks than trains in North America, which logically makes vehicular collisions with Sasquatch more likely than railroad accidents. The first on record, from mid-January 1973, involved a logging truck in Oregon. According to the unnamed trucker:

> I was hauling a load of logs out of Grant's Pass, Oregon, to Eureka. At that time we had to go down the Avenue of Giants. It was about 7:30 in the evening. I was on the highway by myself and had all my lights on...I was doing about 40-45 miles an hour when I went around a curve and as I got approximately to the center of the curve this being, whatever it was, stepped out from the right hand side and I hit it with the truck. The top of the hood is six foot four inches high and the top of its head was about six inches above the top of the hood. I don't know why it didn't see the lights or notice me. I didn't notice whether it was male or female. The upper torso was turned away from me at possibly 45 degrees rotation.
>
> I hit it with the truck and it flew off to the left hand shoulder of the road. I didn't hear any noise it made. Of course, I couldn't hear anything over the exhaust. ...After I hit it I drove down the road about five or six miles and then stopped to check on the damage, which was quite extensive. I didn't see any blood or hair on the front end. The front radiator had been smashed in, away from the fiberglass hood. The glass was cracked in several different places, and we had to replace the whole thing.[13]

When the trucker informed his boss of the incident, he was asked what he'd been drinking.[14]

One year later, on January 9, 1974, motorist Richard Lee Smith reported his collision with a Sasquatch on State Road 27 outside Hollywood, Florida. According to Smith, the creature he struck was "huge, seven or eight feet tall, dark-colored and human-like," but resembled "a gorilla, a strange creature of some sort." It jumped in front of his vehicle, then fled, limping, into the Everglades after he struck it. Police found no trace of the creature, but Smith's story rated an article in the March 1976 issue of *Startling Detective* magazine.[15]

Remaining in Florida, researcher John Green writes that "There was another report of one hit by a car near Gainville [*sic*] the following February [1975]," but he offers no further details.[16] One month later, still in the Sunshine State, Janet and Colin Bord tell us that motorist Steve Humphreys and his wife, driving near Lake Okechobee on March 6, "collided with Bigfoot running fast across [the] road. Much damage to car, but no victim could be found."[17] Further research on those cases for this volume proved fruitless.

Florida got a respite from Sasquatch hit-and-runs on April 28, 1975, when Peter Hureuk of Bel Air, Maryland, met Sasquatch on the highway at 3:00 a.m. According to John Green, Hureuk's statement to police says that he "hit a seven-foot upright creature that crossed the road in front of his sports car. He said he knocked it down but it got up and limped off, still on two legs. The 50-miles-an-hour collision did considerable damage to the car. Police studied some hair taken from the car and announced that it came from a bovine."[18] The Bords place that incident near "Rocks," presumably referring to Rocks State Park, north of Bel Air.[19]

Ray Crowe's IBS database offers our next reported collision, from late January 1977 in Yakima County, Washington, but it includes no details: "TRUCKER HITS BF, DAMAGED TRUCK, BF RAN AWAY." Sadly, the "full report" promised after that teaser does not appear online.[20]

No collision was reported from Ohio's Union County on June 24, 1980, but motorist Donna Riegler may have seen an injured Sasquatch lying on the highway. As she told the *Akron Beacon Journal*—

> I was in a good mood. I just wanted to get home. I went over the railroad tracks slow. I always do because I don't want to knock my wheels out of line. Then I saw this thing laying on the road, hunched over. I thought it was a big dog at first. Then it stood up and I thought it was a man. I thought he was crazy, laying on the road. I couldn't figure why he was out there. He had no golf clubs. No luggage. Then he turned and looked at me.[21]

Riegler distinguished no facial features, and reported no apparent wounds as the creature stood with bent knees and empty hands extended. She raced to a neighbor's home, and on arrival burst into hysterical weeping.[22]

Our next report comes from the area of Chemult, Oregon, where a train allegedly killed one Sasquatch in January 1965. This time, on some unknown date in the winter of 1982, the incident involved a truck. Ray Crowe's IBS database describes the accident as follows (uncorrected):

> Customer Joe Cary used to own a coupla" trucks, one a lumber carrier with a peterbuilt cab. It was winter of 1982, on Hy. 97, just south of the Hy. 58 intersection

(near Chemult, Or.), when Joe"s driver hit a Bigfoot as he was traveling south. He crashed and flipped over the unloaded truck when he hit the 7½ or 8 foot creature on two legs. There was extensive damage to the truck from the BF collision; it took out the radiator and bumper, and a lot of fiberglass stuff. Cary was accused of drinking, being known to be a tippler, but there was no alcohol in his system Joe said. He was also accused of hitting an elk, but Cary said it was on two legs, and taller than the cab of the truck, besides...why would he make up an unbelievable story of hitting a BF when he could have as easily said it was a bear or elk?[23]

Aside from the truck "flipping over," its damages sound remarkably similar to those from the January 1973 Oregon incident, reported above. The confusing business regarding Joe Cary's blood test for alcohol, when another individual was at the wheel, remains unexplained.

In April 2014, while researching *Sasquatch Down*, I posted multiple Facebook requests for any information on apparent manimal killings. The only person to respond was Phil Thackery, whose description of the incident evolved over four days' time. I include it here, unedited, with his permission.

April 18: Hello Michael. Are you still interested in hearing about dead Sasquatches? I don't have time to go into much detail but back in the late 80's I saw something dead on I-81 when they were widening the interstate from 2 to 3 lanes through Bristol Virginia. If you're still interested just let me know and I'll tell what I saw.

April 19: What I saw used to be a forgotten memory. Long story short, I asked myself one day "I've never seen one so why am I helping this guy?" And then it hit me like a ton bricks. I really had to force myself to remember this because I had absolutely t-totally forgot about it. I remember it well now because I concentrated on remembering what I was thinking at the time. I have to go right now but I'll be back to tell you what I saw.

April 20: I'm almost certain it was in early spring of 1989. It was my last year in college at Virginia Highlands Community College in Abingdon, Va. I was either 28 or 29 yo at the time. I lived in Bristol Va. back then and traveled on I-81 everyday to attend class. One Monday morning at approx. 8 am I was traveling north on I-81 and noticed what looked like a person lying face down, arms to it's side, in the southbound emergency lane on the bridge overpass. Exit 5 to be exact. As I approached it I could see a stream of what looked like blood meandering away from the head and pooling a few feet away. I thought "OMG, someone got hit, or shot and dumped, last night!" As I got parallel to it I noticed it was covered in hair about 3 inches long. The color of the hair was blonde with a reddish tint to it. It was of average human height and dimension. I thought "good grief what is that?" Is that what people are calling a Bigfoot? I've never heard of them being around here, only in the PNW. Beside it's to small and the hair is the wrong color." Anyways, I settled on it was a person with a genetic defect. I wanted to stop and get a better look but it's illegal to stop on a bridge overpass and I thought at the time, knowing my luck, the police would roll up on me when I was stopped and give me a ticket. And if it was a person who was the victim of foul play I would become suspect number one. I figured that whatever it was would be gone by the next morning anyways. Next morning it was still there. It had started raining that day and the hair was matted down and a dirty tan in color where the cars sprayed road grime all over it. But it's shape remained the same. It looked just like a person covered in hair lying there. As the days ticked by I started wondering to myself "are they ever going to remove that poor soul?" I was on my way to work the following Sunday night and saw a state trooper parked on the bridge overpass behind it. He had his interior light on and I could see he was filling out paper work on it. Next morning (yep, after seven days) it was finally gone and I was relieved. I don't know for a fact what it was. At the time I was just glad it was finally gone. Glad it was gone because it was tempting to do something I could potentially get in trouble over. I then proceeded to totally forget about it. Knowing what I know now, I'd give it a 95% chance that it was a Bigfoot that got hit and killed. Back then I'd say no one knew what it was, (myself included) so didn't want to get involved and report it.

April 21: Well no one reported this, poor thing layed there for 7 days. If it weren't in such a difficult spot I would have stopped to check it out, because I really wanted to. I didn't know much about Bigfoot at the time but this thing had my curiosity meter pegging off the charts. If you want to check out where it was, go to Google street view, Bristol Virginia, I-81 exit 7 and head south on I-81. After you see the exit 5 off ramp you'll come to the bridge overpass. It was face down on the southern (far) side of bridge/overpass about 5 or 6 feet from the end from what I remember. I was in a bad spot for sure.[24]

If accurate, this case clearly describes a dead Sasquatch, but the eventual vanishing of its remains renders proof—and scientific analysis—impossible. What became of the carcass—and how police failed to notice it for a full week—remains unexplained.

Our last case with a date attached reportedly occurred during August 2001, near Grants Pass in Oregon's Josephine County. The unnamed driver was a contract deliveryman for an auto parts distributor, driving south on Interstate 5 near the exit for Merlin, towing a 16-foot trailer behind his Dodge pickup truck. He writes:

As was typical for a weeknight at 1 o'clock in the morning there was nearly no other traffic on the road with me. With my cruise control set at 73 mph my headlights lit up something laying centered in the right hand lane directly in front of me. I knew there was going to be contact but rather than swerve and risk lossing control I chose to line up and attempt to strike the object with the undercarriage of my truck. I don't remember touching the brakes or making any attempt to slow down. In the seconds leading up to impact all I could do was brace myself and wonder what it was that I was about to splatter down the highway.

Driving as many miles as I did I had, or thought I had, seen every form of indiginous wildlife Oregon has to offer. Either alive and scampering or squished beside the road I didn't need to see more that a glimpse of fur to identify a critter, even at night. As I closed on this object I couldn't identify it. It had the coloration of deer, but it was much bigger. It wasn't an elk as I didn't see any legs or the characteristic outline of the hip, shoulders and tapered neck had it been laying the other way. This was fast turning from an object to a body. A large, hairy body laying in the fetal position with it's back facing me. As weird as this was it was about to get even more interesting...

Confused with what I was seeing and braced for what was going to be a bumpy ride 'something' flashed directly in front of my bumper from left to right. As close as this thing was to my truck all I could see was the flash of brownish gray hair as it crossed in front of me. It was like sitting, parked in a car, at night and someone walks from one side of the car to the other. All you see is the strobe effect as they pass in front of the headlights. But I was moving at 70 mph!

I'm 5'10" tall and if I stood next to the headlights on my truck my shoulders are at the height of the headlights. What I saw was more the rib section of the creature that crossed in front of me.

Whatever this was it was big enough, strong enough, fast enough and felt the need to pull a 600 pound creature out of the way of my truck because not only did I miss what ran in front of me, I didn't hit anything![25]

Was this a case of one Sasquatch trying to rescue another? If so, who struck the creature lying on the road when the trucker arrived? As usual, we are left with tantalizing questions, forever unanswered.

The IBS database online provides teasers for two Washington cases, but as usual for items transcribed from the Evergreen State, neither is accompanied by the promised "full report." The first, from King County, credits witness Sue Miller with sighting a "BIGFOOT ROAD KILL." The other, logged from Kitsap County by "Rhonda and her mother," reads simply "BF HIT BY CAR, SCREAM."[26]

Ray Crowe closes this chapter with nine undated incidents, dropping incomplete names that render the stories untraceable. Those accounts, presented here as they were posted to the Web in March 2011, read:

Leroy heard on his scanner that on Highway 35 close to Odell Junction near Hood River, Oregon, a cop had hit a Bigfoot. It totaled the car and the cop was consigned to the hospital.

Craig and wife were on I-84 one half mile west of Cascade Locks, Oregon, when they saw a grey Bigfoot body in a fetal position off the side of the road.

Again Clarisa and her brother were on I-84 thirteen miles east of Hood River, Oregon, and saw a grey Bigfoot body, again in a fetal position only it had a long stretched out arm. The corpse was lying next to the freeway, she was sure it was not a deer or elk, and it was in the westbound lane. Sounds similar to the previous report.

A year before, my wife Theata and I investigated a similar report from the same area, but could find nothing but Bigfoot tracks on the riverbank. Was this a road kill and somebody or something is picking up the bodies? Nothing is ever seen in the papers. You never hear anything about a probably very interesting news item—conspiracy afoot? Papers seem more interested in a good hoax though.

Near Willows, California, Chambers hit a Bigfoot at Elk Creek on Highway 261 off I-5. The ten foot animal was gone when he returned two hours later to check. Survived? Maybe also carried away by a grieving mate?

Bob's wife saw in a wooded area on Route-95 in Hartford County, Maryland, a hair covered human being lying in the road as they passed around the emergency vehicles.

Rob called the Art Bell Show and said that he had hit a 700# BF near Ironwood, Michigan, which was squatting in the road while he was doing 70-75 mph (musta' given him some dents, car and his head).

A logger was barreling off a mountain road with a load of logs, and hit a Bigfoot with the right front side of his Peterbuilt, throwing the ten foot Bigfoot off the road into a ditch. Two days later the body was gone with only some blood and hair remaining. It is pretty well known that the carcass of an elk or bear can be completely gone by scavengers within 24 hours.

A BF was hit and killed near Yale, Washington. A County Roads Department worker was left to block the road until a Forest Service chopper picked the body up. No more details, of course.[27]

One of those incidents receives more detailed coverage in the IBS sightings database. According to that report, "Clarisa and her brother" logged their sighting around 5:00 a.m. on July 2, 2000, in Oregon's Hood River County. Clarisa was "slightly irritated" when researcher Steve Williams inquired whether the creature might have been a deer or elk, "as she was very familiar with Oregon animals."[28] Williams wrote:

They were about 13 miles east of Hood River on a straight stretch of road when they noticed a large gray animal lying in the middle of the road, and they slowed down to a near stop to look. At first they thought it might be a bear. It was lying in a fetal position with what Clarisa describes as a very, very, long arm stretched out. When they realized they might be looking at a Bigfoot, Clarisa wanted to stop and look further but her brother got scared and wouldn't let her out of the car. It was lying in the west bound lanes but when another relative came by about an hour later, it had somehow got over into the east bound lanes (same thing?). That part of the freeway has a cement wall in the center...when she later called her brother, he didn't want to tell anyone. He had told his wife and she had laughed at him.[29]

Williams visited the site and found nothing but a road-killed porcupine. Ray Crowe and his wife followed up sometime later, scanning both shoulders of the highway, and found nothing. Crowe wrote, in conclusion, "Don't know what happened to body if there was a road kill...recovered, another creature carried it off, or, of course, a possible misidentification."[30]

And the rest, at least for now, is silence.

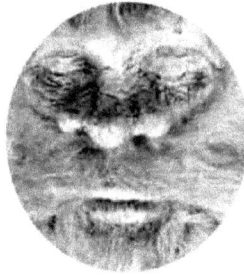

Chapter 7.
Buried Secrets

S ome Sasquatch believers explain the scarcity of manimal remains by suggesting that the creatures live in families or tribes, claiming the corpses of their dead for burial whenever possible. We have already mentioned several cases wherein dead or badly wounded Sasquatches vanished from a shooting scene, and the Oregon trucker's tale from August 2001 suggests that corpse retrieval may be carried out even at risk to life and limb. From those anecdotes, some writers have advanced to taking Sasquatch burials as an article of faith.

Blogger Sally Lou Hock writes: "Bigfoots live in family groups. Where you see one, there are others about. There is also evidence that Bigfoots bury their dead."[1]

The BFRO's website says: "It is quite possible they bury their dead; they may have learned this by watching First Nations People and the first white settlers to North America burying their dead."[2]

Ray Crowe asks, "Perhaps the Bigfoot buries their dead companions, as was suggested earlier, even with reverence, with perhaps a 'religious' rite of some kind?"[3] To which blog reader and Sasquatch eyewitness Dr. Rhettman Mullis Jr. replies categorically, "Bigfoot buries their kind, when possible."[4]

Robert Lindsay states with conviction that "From 1949-present, a 62 year period, 14 Bigfoot burials or possible Bigfoot graveyards have been seen. Therefore, Bigfoot burials or graveyards are seen...about once every 4½ years."[5]

But how reliable are those reports?

John Green logged the earliest account of a Sasquatch "funeral," then dismissed it as untrue, writing: "In the 1950's I had a letter from a man named Peter James, who claimed that in 1949 he saw two old males and a female sasquatch lay out the body of a dead young female on a rock high on a mountain. I understand he died shortly afterward. In any event I never tried to talk to him, mainly because I didn't put any stock in it."[6] Ray Crowe laments that lapse, writing that the "sky burial" tale "seems reasonable in light of some of the following reports."[7]

Next in line from Crowe is this account: "[Roger] Patterson said a witness saw three Bigfoot burying a fourth, digging the grave with their hands. Perhaps it was to keep predators from eating them? Then they piled rocks on the grave."[8] While Crowe provides no further details, Robert Lindsay places the event in northern Washington State, dating from the period of 1962-67, but credits the story to Peter Byrne.[9]

Lindsay dates our next account from sometime "after 1972," while Ray Crowe provides the best report, based on communication with the witness. He writes:

> Vic McDaniel reported that a relative near Klamath Agency, OR, while constructing a road, ran his [bull]dozer through a curious twenty foot ring of large boulders, with smaller stones in the center. The next day and on several peculiar occasions afterwards, the stones had been replaced in their original positions after he had 'dozed them away. He eventually built his road around the area. Don't have a specific Bigfoot in this tale, but speculate that the stones were moved by something big and strong...perhaps a Bigfoot grave? How long would a Bigfoot mourn a dead companion? Or perhaps, a Bigfoot was playing some complicated game with one of those "curious" humans.[10]

In the Pits
Our next report is confused and plagued by contradictions. Ray Crowe was first to report it, in March 2011, writing: "Again Peter Byrne related in a recent telephone conversation, where he had a new report from an elderly gentleman; that 35-40 years ago [i.e. 1971-76] in northern Washington (still being investigated), the witness had seen three Bigfeet burying a fourth. With the multiple numbers of animals reported in the burial rites (usually three or four), one could almost speculate on a 'family funeral service' of intelligent beings. Can they be so much different than ourselves? And so we suspect where the bones might be?"[11]

That appears to be a retread of the tale mentioned above, first credited by Crowe to Peter Byrne. Crowe's IBS database files are rife with duplicate reports, and even triplicates listed under different case numbers, but this one spawned further confusion.

Robert Lindsay chimed in next, saying: "Before 1975: Location unknown, probably Pacific Northwest. Three Bigfoots were witnessed digging a hole with their hands to bury a fourth Bigfoot. When the hole was filled in, huge boulders were rolled over the site. Reported by Glen Thomas."[12]

While we would normally appreciate the witness being named, this time the I.D. creates more problems than it solves. Researcher John Bindernagel introduces Glen Thomas briefly, writing: "In October 1967, in a remote part of Oregon's Cascade Range, Glen Thomas observed a male sasquatch dig a pit, from which it obtained and ate ground squirrels."[13] Another Sasquatch hunter, Craig Woolheater, adds more details to the story.

> Glen Thomas, a logger living in Colton, Oregon, eventually claimed four separate sightings—which is more than enough to set alarm bells ringing—but his first story, of watching a big male sasquatch dig deep into broken rock high up on a mountain ridge to get at hibernating rodents, was backed up by the hole in the rocks, five feet deep, as steep-sided as a well, and obviously beyond human ability to duplicate without machinery. He also had something to say bearing on the family hypothesis. A female and infant were with the big male and shared in eating the rodents, but Glen noted that the young one was always careful to keep on the other side of its mother from the male.[14]

Lindsay has apparently confused one sighting with another, naming Thomas as a witness to a Sasquatch burial, when in fact he saw nothing of the kind. James Hewkin, from Oregon's Department of Fish and Wildlife, confirmed as much in June 1992, when he examined "the Glen Thomas rock pit" with John Green and researcher Jack Sullivan. Hewkin reports:

> The purpose of the trip was to record measurements of the pit and the rocks. The pit diameter was measured at 7 to 8 feet across the top, and it tapered to 3.5 feet (1 m) across the bottom. The depth was 5 feet (1.5 m). It should be noted that 25 years had elapsed since the pit was excavated, so there had been some natural displacement of rock. Indeed, on a later trip to the site in October, the pit had caved in considerably on one side, causing a wider configuration and shallower depth.

> We weighed the rocks on bathroom scales—with some difficulty due to hazardous footing on loose, irregular shaped rocks. Of the seven rocks that were weighed, the smallest was 35 lb and the largest was 240 lb...

> Because of footing difficulties, it required two men to handle the rocks. In reflecting back on Glen Thomas' account, it is evident that great force would have been required to both free these rocks from crevasses and fissures and to lift them out. The strength of the animal involved had to be phenomenal. Measurements of a more recently dug pit, about 30 yards (27 m) from the aforementioned pit, indicate a depth of 3 feet (91 cm) and a 3-foot (91 cm) diameter. Of interest is the observation that it was dug after 1973, the year that I first visited the site and noted only a few rocks pulled out. By the visible slight weathering, this pit appears to be 10 to 16 years old.

> In regards to Thomas' observations, one might propose quick success by these animals in locating hibernating rodents, and that this behavior has been established over a long period of time. Credence is given to this proposition by information regarding similar pits located in the Gifford Pinchot National Forest, in the state of Washington.[15]

Clearly, the team was not searching for Sasquatch remains, and indeed found none. Whoever claimed the 1970s burial sighting from Washington, he remains anonymous today. And yet, the garbled tale persists, with Ray Crowe making matters worse, turning one event into two. He writes:

> Thomas described where three Bigfoot adults dug a deep hole with their hands, and interred a fourth dead Bigfoot. After the hole was filled in, they rolled some huge several-hundred-pound boulders over the area. Peter Byrne also said an elderly man in northern Washington saw three Bigfoot burying a fourth.[16]

Then, he seems to merge the two stories: "Peter Byrne quotes of a report from Glenn [*sic*] Thomas ('Bigfoot,' 1975, Guenette, and 'Bigfoot: Opposing Viewpoints,' 1989, Gaffron,) describing where three adult Bigfeet had dug a deep hole with their hands as tools, and buried a fourth dead Bigfoot. After the hole was filled in they rolled huge boulders, weighing several hundred pounds each, over the site."[17]

In my research for *Sasquatch Down*, I obtained copies of Norma Gaffron's *Bigfoot* (a children's book published in 1988, not 1989), and *Bigfoot: The Mysterious Monster*, written by Robert and Francis Guenette as a companion volume to their 1975 film, *The Mysterious Monsters*. Gaffron's volume offers the following tidbit, with no mention of Glen Thomas.

> Byrne says that perhaps Sasquatches "bury their dead and bury them deep." He tells of a man who swore he watched three Bigfeet burying a fourth. After digging a deep hole, using only their hands as tools, they placed the body in the hole and covered it with earth. Then they rolled huge boulders onto the grave. This grave has never been found.[18]

The Guenette tome also quotes Byrne as saying that he "came across an account where an eyewitness declared that he actually saw several Bigfoot creatures burying a dead fellow member of their species," but again, there is no mention of Glen Thomas.[19] There ends the thread of evidence, hopelessly snarled beyond untangling.

Graves in Waiting?

Ray Crowe reports the next discovery of a purported Sasquatch burial ground in Linn County, Oregon. I have corrected his misspelling of the Calapooia River, a 72-mile tributary of the Willamette, but otherwise leave his account as posted to the Web.

> Strange piles of rocks, possible burials, in the forest are common in issues of the Track Record. One in particular tells of gold miners finding in 1985, a fifty foot clearing in the forest near the Calapooia River, OR, where the tops of all the trees around the perimeter were broken off. In the clearing also was an old deer carcass and two piles of smooth, five-inch cobble rocks, about two and a half feet high, and separated by twenty feet. The miners had been scared the previous evening from strange screams and the sound of breaking trees. Was never able to excavate these particular piles, but am sure others are out there waiting.[20]

Robert Lindsay reports a similar find, four years later: "1989: McMinnville, Starkey, Oregon. Scott White found strange piles of rocks in a clearing with smashed trees. Possible Bigfoot burial site."[21]

The following year brought reports of an actual corpse being found. Lindsay writes:

> 1990: Estacada, Oregon. East of Portland, a hunter found a dead baby Bigfoot ten feet up in a tree. He reported that it was just a small, furry little thing. He was interested in the tree in the first place due to large scat piles all around it. The dead Bigfoot was buried in the boughs of an evergreen and was covered with other boughs. The hunter thought that the scat piles were from the mother Bigfoot who had been sitting under the tree mourning the death of her baby. This could be called an "Indian style burial," as Indians in the Pacific Northwest used to bury their dead up in trees, albeit in caskets. The hunter called Portland State University and told them he had found a dead baby Bigfoot. They laughed at him and told him that they were not interested in looking at it. After all, Bigfoots don't exist. The man stuck the baby Bigfoot in his deep freeze, and that's the last we've heard of it. Reported by Ray Nab.[22]

Ray Crowe's IBS database provides a few more details.

> Called in by Ray Nab, an electrical contractor; the secondhand tale went: Near Estacada, OR, a Molalla hunting friend said he had found two years ago, a dead...baby Bigfoot, a brown female, "just a furry little animal," that was left ten feet up in the boughs of a tree, and then was covered with other boughs...an "Indian style burial," he said. (is it possible that it was?) Attention was called to the tree by piles of huge droppings around its base. One can almost visualize a grieving female Bigfoot staying near the remains of her dead young, as has been reported with many other primates (snow monkeys and gorillas)...have speculated on disease causing the death of the infant. Portland State University was contacted, the finder thinking they would be very interested in a new species of animal. He, of course, got the old horse laugh...they wouldn't even look, and the creature reportedly is now in his deep freeze. The college later called back, apologized, and asked to see the corpse, but the harm had been done, and he told 'em where to go. Still, even this tree "interment" is again quite simple, and might have been done by a creature without any "intelligence," in the human context...elephant like, if you will. Names and addresses have been released to Peter Byrne for further investigation. If not a hoax. He hopes to procure tissue samples for DNA analysis. He also says any information will be shared with the WBS. Peter also comments he has found human Native American tree burials, though in coffins, in B.C., Canada (The witness ultimately disappeared, and contact was not made).[23]

Estacada produced another tale of Sasquatch burial two years later. Robert Lindsay writes: "Summer 1992: Estacada, Oregon, near Bagby Hot Springs. A philosophy teacher saw two Bigfoots, either a male and female or two females. There were two young, auburn colored Bigfoots with them. They were in a riverbed, burying another Bigfoot under a pile of stones. They had not dug a hole; they were just burying it with rocks. He stated that the Bigfoots were acting 'sad.' The site was rechecked by an investigator one year later, but flooding had washed the stones away, and the site could not be rediscovered. Reported by Ray Crowe."[24]

Crowe elaborated in his March 2011 blog.

> Considerably more advanced in technique, is another new report from the Estacada area in the summer of 1992. The investigator wants to sit on this one until he checks it out personally, but gave me the essentials. His informant, well educated...teaches philosophy, had an experience east of Estacada last summer. He was hiking, 4-5 hours into the woods, when he heard a, "clink, clink, clink," of rocks. Curiously approaching the upstream noise, he was less than a hundred yards away from two Bigfeet...he wasn't sure if the larger was a male and the smaller a female, or two females. There were also two smaller red colored young Bigfootlets. All were engaged in burying another dead Bigfoot under a pile of stones. They had not dug a hole, but were just covering the body with the stones, causing the "clacking" noise. The informant said he had an intense feeling of sorrow, and that the Bigfeet were acting "sad." (Realize that I might be somewhat anthropomorphic in reporting that piece of information.) Maybe we'll have some bones from this one this spring? But no, later the investigator found that flooding had completely washed away the stones and he couldn't find the site again.[25]

Flash forward to our next account, from Washington State. Robert Lindsay reports: "After 1995: Whidbey Island, Washington. Rhett Mullis found large mounds on this island in Puget Sound where there is no history of Indian residence. A large pit had been dug out but had not yet been used. There was a 'hallway' along a well-used trail and scat was scattered around. The mounds were covered with large hand-sized rocks. Plants had been pulled up and placed on top of the mounds in order to hide them. Possible Bigfoot graveyard."[26]

"Possible," of course, proves nothing. Mullis notes that Whidbey Island has no history of indigenous people to account for supposed burial mounds, and goes on to say—

> My first sighting was of a Bigfoot swimming from the Olympic Peninsula to Whidbey Island. Perhaps these are bodies of Bigfoot that did not survive the swim since the area is located on the top of a bluff in an old-growth forest very near the beach on the Olympic Peninsula side of the island. Because these mounds and others like them are surrounded by track evidence, I also found a pit that was dug out but not used yet which correlates to one of the above reports, as well as feces and a "hallway" formed along a well-used trail leading down to the beach. The mounds were covered with large hand sized rocks, perhaps carried as they traveled through. The mounds were also strategically covered with limbs and plants pulled up out of the ground and placed there to hide the mounds. A deliberate and cognitive act.[27]

It is regrettable that no one ever seems to undertake an excavation of such mounds—or, in some cases, even photograph them or commit their location to memory. At the very least, if those who find the sites lack opportunity or interest enough to delve beneath the surface, someone else might take them up on it.

Our next case takes us to Ohio, scene of 253 Sasquatch sightings according to the BFRO's website. In May 1999 the *Portsmouth Daily Times* published an interview with researcher Dallas Gilbert, who claimed nine personal manimal sightings since 1978. Along with close encounters, Gilbert described a monument of sorts in Scioto County. "It looks like a tombstone almost," he told reporter Kirsten Stanley. "You can see the outlines of the creature's eyes, head and his teeth. There's nothing else like it in the world, from what I know." Stanley recorded Gilbert's claim that he had discussed his findings with a primatologist in Columbus, Ohio, but she noted that "he would not give the expert's name or what she said about his alleged discovery."[28]

Autumn of 2002 takes us back once again to the neighborhood of Estacada, Oregon. Robert Lindsay writes: "October 21, 2002: Estacada, Oregon. Possible Bigfoot burial grounds consisting of pits and stacks of heavy rocks were found at a high elevation in the Clackamas River Gorge. They could not be Indian burial grounds. Reported by the BFRO."[29]

In fact, researchers Joe Beelart and Richard Noll made the discovery, near "the famous Glen Thomas site." As Noll informed the BFRO, "This evidence is in the form of 6' deep pits dug into these rock screes and the removed 100lb+ rocks neatly stacked nearby.... Joe told me that further up the hill, in other rocky areas, there may be as many as a hundred of these pits and rock cairns. They had also found things that looked like long mounds in these rocks, thinking it might be a burial site. He contacted local authorities and was told not to mess with them. Local Indian archeological sites have nothing that looks like these pits and cairns. They would not build anything at that altitude nor that exposed to the weather. The high side of the pits I saw were on the southern side." Once again, no excavation was attempted, and while Noll suggested returning with fiber-optic cameras to probe the mounds, nothing was done in that regard.[30]

Our last report bearing a date was posted online on April 6, 2014, under the headline "Watch: Park Ranger Witnesses Bigfoot Burial [Real Bigfoot Encounters]." A teaser tells prospective viewers, "This report comes from the High Sierra in California. A forest ranger witnesses a 'Bigfoot burial' after a camp was ransacked by a clan of Bigfoots." Alas, pressing the video clip's "PLAY" button produces nothing but a snowy screen behind a curt message reading "This video is private."[31]

Undated Offerings
Alleged observations of Sasquatch burials logged without dates further confuse our narrative. Ray Crowe's IBS database leads off with a terse item reading "BF FAMILY BURIES BF UNDER ROCK," credited to researcher Robert W. Morgan, but the promised "full report" does not exist online.[32]

Next up, Robert Lindsay writes: "Date unknown, modern era: Northern California. A witness saw four Bigfoots carrying bones. The longest bone was up to four feet long. Reported by Ray

Crowe."[33] Crowe's online database summarizes that case with the notation "GIANTS BURIAL PLACE," while omitting the "full report" promised[34], but he did mention the incident in a March 2011 blog. There, he wrote, "Carene [?] reports that a witness saw a small group of Bigfoot carrying huge bones in Northern, California. The four creatures walked in single file with a huge four foot bleached white bone balanced over its shoulder. The others also carried bones, but they were smaller. Four feet?"[35]

Lindsay logs our next report: "Date unknown, modern era: Starkey, Oregon. In the Blue Mountains, Sue Sebring found unusual cobble piles in the forest. Possible Bigfoot graveyard. Reported by Ray Crowe."[36] Crowe's passing mention of the incident, posted two months earlier, adds no substantive details.[37]

Lindsay's next item is equally vague: "Date unknown, modern era: Alder Creek, Sandy, Oregon. East of Portland, Peter Byrne noted an unusual mound of earth along the creek."[38] Ray Crowe offers a similar frustrating story: "Chuck [?] says he was on the lower Roaring River near Estacada, Oregon, when he found a five foot long earthen mound at water's edge that was overgrown with weeds. He dug in it for a short distance, but found nothing."[39]

Finally, the Sasquatch Tracker website brings us full circle with a report reading:

> Unknown/04/L –Klinkwan, AK
> Alleged burial site near Klakas Inlet, Prince of Wales Island.
> IBS #145, 2248[40]

This is, in fact, the same turn-of-the-century shooting previously covered in Chapter 5, although Sasquatch Tracker omits the reported killing. It also confuses its sources, since the IBS database contains *three* separate listings for this alleged incident, confounding the matter further. Each file offers geographic coordinates, but none agree with one another. Report #145 places the site at Latitude +054° 59' N, Longitude 132° 25' W; Report #2248 pegs it as Latitude +055° 10' N, Longitude: 132° 35' W; and Report #3342 puts the gravesite at Latitude +055° 38' N, Longitude: 132° 54' W—all three locations supposedly calculated from the same "old map" supplied by Peter Byrne.[41]

If this were not confusing enough, Crowe cannot even decide how much he trusts each item. On a "credibility" scale ranging from 1 to 5, with 1 rated most likely to be true, Report #145 rates a 4, while Reports #2248 and 3342 each rate a 3.[42]

Before we close the door for good on Sasquatch graves, however, there is more to learn.

. Daniel Boone—first North American Sasquatch slayer?

Anonymous photo of an alleged Sasquatch killed in 1894, location unknown.

Hoaxed Internet photo of a backwoods hunter with his bipedal kill.

Hoaxed Internet close-up photo of a supposed dead Sasquatch.

A typical U.S. tabloid spoof report of police killing a Sasquatch.

YOUR NOSE REVEALS EXACTLY HOW LONG **YOU WILL LIVE!**

WEEKLY WORLD NEWS

AMERICA'S EXTREME NEWSPAPER

BIGFOOT BABY FOUND

...abandoned outside Neverland Ranch!

Would YOU adopt this cuddly furball?

There's gonna be one helluva custody battle

The same tabloid claims a live Sasquatch capture.

Another anonymous Internet photo purporting to show a slain Sasquatch.

Contemporary report of the alleged Ape Canyon battle, 1924.

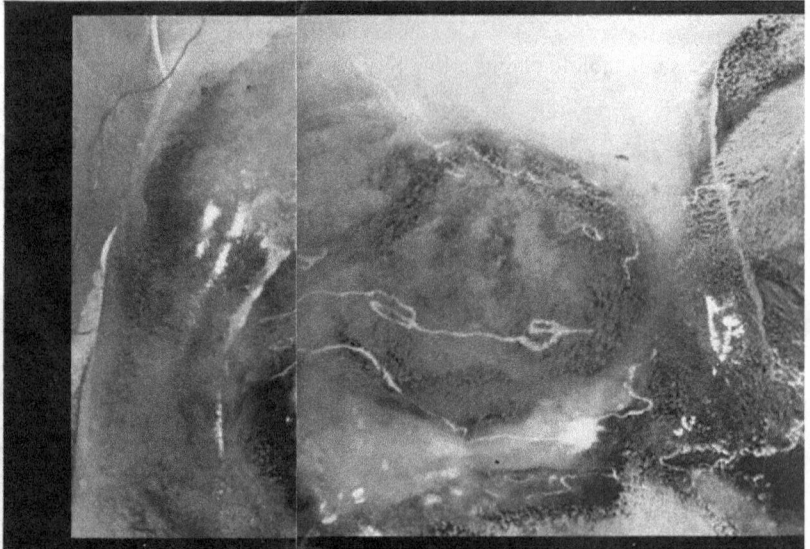

Saga magazine article on the Minnesota "Iceman."

Roger Patterson's "Patty," alleged victim of a wholesale Sasquatch massacre in 1967.

BIGFOOT TIMES

All Rights Reserved, Copyright © Daniel Perez

10926 Milano Avenue, Norwalk, California 90650-1638
www.bigfoottimes.net

December 2007

Bigfooter Of The Year: M.K. Davis

About ten years ago a Benton, Mississippi man, Marlon Keith (MK) Davis, 51, viewed some cibachrome prints of the best frames from the P-G film and was greatly impressed.

"Fabulous," he thought to himself about the cibachrome work done by Bruce Bonney on the best frames from the famous 16-millimeter Bigfoot movie.

Said MK Davis in a recent interview, "there is nothing you can do to rescue a good photo from a bad film," and the P-G film, in his estimation, is quite "good." The work done by Bruce Bonney using the "most aesthetic looking frames" from the movie with a process known as "glass filtration" was "very, very good."

Cibachrome is now widely known as Ilfochrome and is a positive-to-positive photographic process which tends to bring out clarity, color purity and super saturations of the colors. When Bruce Bonney did this work in 1980 with the late Rene Dahinden - what was crucial - was having access to the *original* movie film. At the time, this was considered "cutting edge" in the then small, snail mail Bigfoot community.

The Bonney-Dahinden work clearly demonstrated that key frames from the P-G footage were "oh my god" sharp and clear, in spite of what you may have heard to the contrary.

In that ten year time span, MK Davis would become a student of the P-G film, rising to the top of the class.

"The little light came on," Mr. Davis said, as he launched his own photographic and computer research of the film using his background in astrophotography to lead him in different directions.

Astronomy buff, M.K. Davis seen at home with his prized telescope. Center, Davis in the Bluff Creek, California area on a research trip and below, lecturing in Texas on the finer points of the subject in the P-G Bigfoot film. Photo courtesy of M.K. Davis. Bottom photo courtesy Chester Moore, Jr.

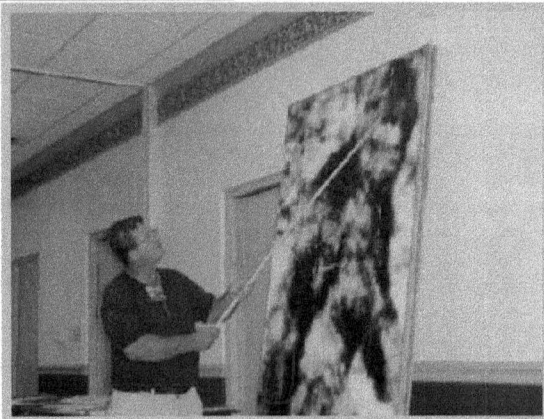

1

Bigfoot Times article on M. K. Davis, originator of the "Bluff Creek massacre" story.

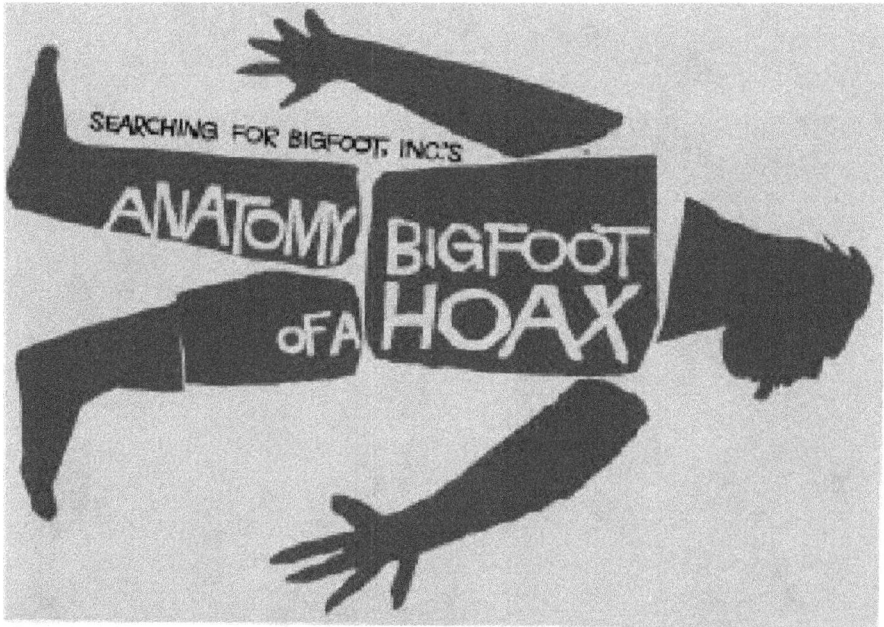

SEARCHING FOR BIGFOOT, INC'S

ANATOMY of A **BIGFOOT HOAX**

TOM BISCARDI
BOB SCHMALZBACH
REX HOWDYSHELL
ROBERT BARROWS
WILLIAM LETT
TJ BISCARDI
JC JOHNSON
LEONARD DAN
(the victims)

with MATT WHITTON & RICK DYER (the hoaxers)

SEARCHING FOR BIGFOOT, INC presents the unedited uncut video revealing the ANATOMY of a BIGFOOT HOAX as told by the SFBI Team with actual scenes from the hoax in the order that they happened.

www.searchingforbigfoot.com

Poster for Tom Biscardi's film on the 2008 Georgia Sasquatch hoax.

In October of 2010, 2 young hunters shot a 7 foot tall animal they could not identify. It stood upright and walked like a man.

DEAD BIGFOOT
A TRUE STORY

MULDERSWORLD.com presents
a RO SAHEBI FILM · DEAD BIGFOOT: A TRUE STORY
JUSTIN SMEJA · BART CUTINO · RO SAHEBI · SANH ORIYAVONG
Written, Produced & Directed by RO SAHEBI · Bigfoot suit by DOUG HUDSON & KLONEFX.com
Produced in cooperation with SIERRAS EVIDENCE INITIATIVE · DEADBIGFOOT.com

Publicity release for Justin Smeja's film touting the "Sierra kills."

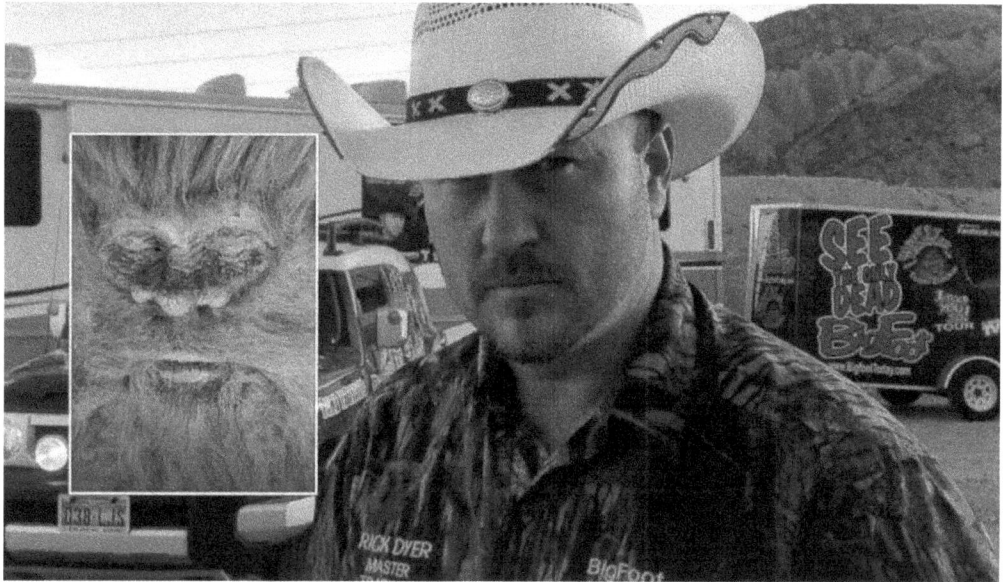

Rick Dyer with the second incarnation of his dead Sasquatch

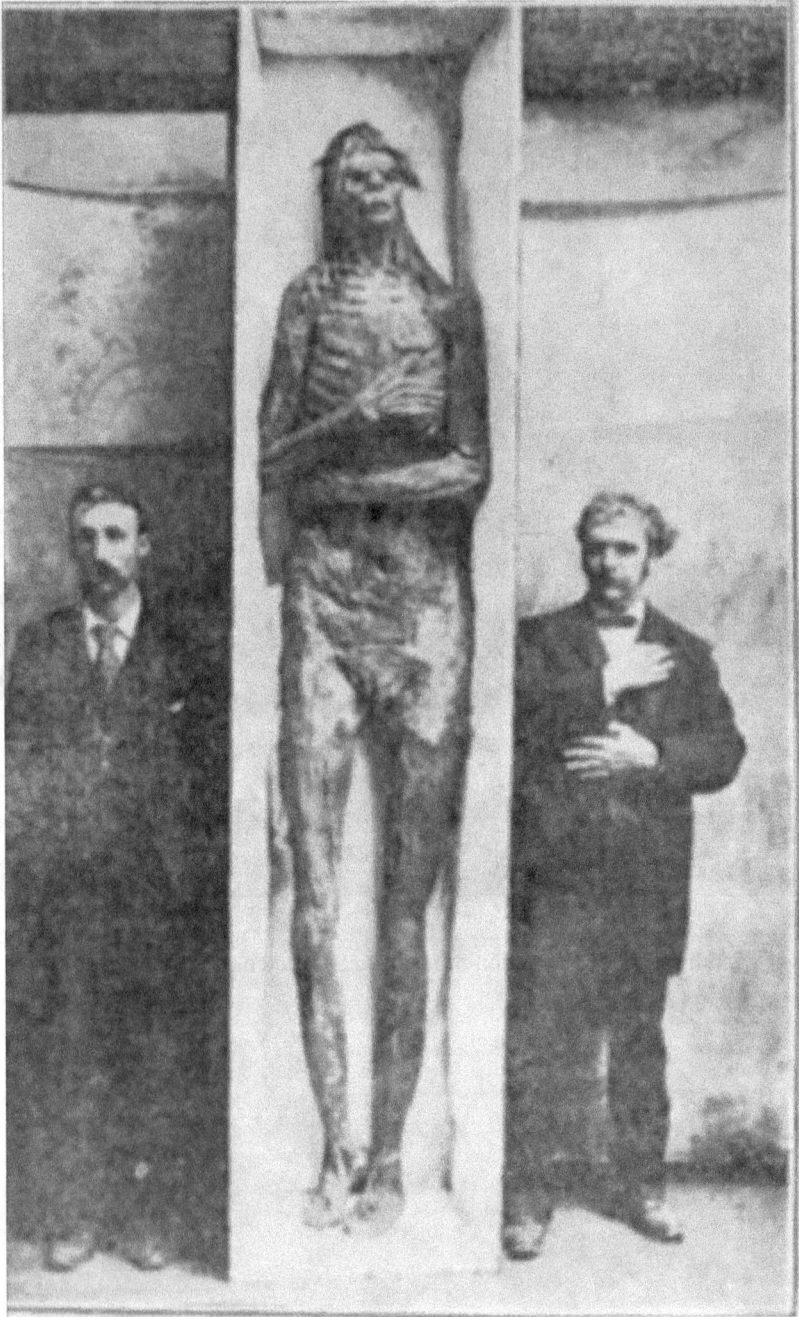
Purported remains of a giant unearthed in the 19th century.

THE CLEVELAND LEADER, MONDAY MORNING. AUGUST 11, 1884.

A WESTERN WHAT IS IT?

A Mysterious Creature in British Columbia.

The village of Yale, B. C., is situated at the head of navigation on Fraser River, ninety miles above New Westminster, which was the capital of British Columbia until it was changed to Victoria. About twenty miles from Yale, on the line of the railroad, is a locality roughly known as "Tunnel No. 4," where the extraordinary occurrence about to be related took place during the early part of the present month.

Notwithstanding the improbability of any amount of prospecting resulting in turning up even the bones of the "missing link," much less in finding an actual living specimen of this much debated being, the actual facts which are related concerning the remarkable appearance near "Tunnel No. 4" would tend to bear out this theory of the subject. At different times during the past two years there has been seen in the hilly country about the settlements a being whose personal appearance is variously described.

One day about a year ago a party of young people from Yale went up on the road as far as Tunnel No. 4, and there, disembarking from the cars, proceeded to spread themselves over the country in the form of a picnic party. The tempting meal had been spread upon the ground, and young men and girls were seated in a circle preparing to enjoy the viands, when there was heard a loud crashing noise above their heads, and in an instant, without further warning than was given by a most fiendish yell—something between the shriek of a hyena and the Indian war whoop—there dropped into the midst of the spread a horrible creature as large as a man, covered with hair from head to foot, with long arms which he brandished about in formidable style, as he vainly tried to extricate himself from the canned fruits, cold meats, jam pots, and oleomargarine into which he had unexpectedly tumbled. This was a "surprise party" for which no intentional preparation had been made, and in a moment there was a stampede.

Tumbling headlong down the hill on whose crest the elaborate meal had been laid, the frightened picknickers so hastened their departure as to be utterly unable to give any coherent description of what had frightened them to the railroad men whose assistance they implored. A party fully armed was at once made up, and the scene of the sudden onslaught was carefully approached. The unwelcome visitor had fled, but before leaving he had plainly helped himself to everything that took his fancy, and that seemed to have been guided by nothing but the opportunity. If he were a human creature and had eaten what was certainly gone, selected from every imaginable article of food, his remains would undoubtedly be found in a few hours. No idiot Indian or other kind of man could possibly have eaten such a mixture and live.

But if such was the case, the most careful search failed to result in finding the body, and after a protracted search, which lasted, after a desultory fashion, for several weeks, the idea of his having died of indigestion or gout was reluctantly abandoned. One fact which was demonstrated by the circumstances of this visitation caused the believers in the Indian theory to be very deeply shaken in their convictions. This was that he had fallen from an overhanging limb of a tree, carrying a large piece with him, and the size of the limb was a good indication that the creature must be as heavy as an ordinary sized man, and hardly an Indian, as they do not usually climb trees. A few months later another view of this strange being was had by some workmen on the railroad, but, though they gave chase, they were not able to come up with him. He was not seen again until about three weeks ago, when he was not only seen, but caught. The spot where he was discovered was a series of bluffs, deemed inaccessible. A train was running from Lytton to Yale, when the engineer saw what he supposed to be a man lying close to the track. He whistled down brakes, but just as the train stopped the object sprang to its feet, and in an instant the object was climbing the side of the precipitous declivity with the greatest ease. The conductor, brakemen, express messenger, and a number of passengers at once gave chase, and after some perilous climbing succeeded in corralling the creature on an overhanging shelf of rock from which he could neither ascend nor descend. The ingenious, though rather cruel, method was now adopted for securing him, of dropping a piece of stone from above, which, falling on his head, stunned him, and he fell insensible.

The bell rope was now procured, and, after some expert climbing, he was reached, tied, and lowered gradually down to the foot of the cliff. He was placed in the baggage car and successfully transported to Yale, when it was found that he had recovered from his insensibility, and was tractable and docile. One of the men in the railroad machine shop assumed the care of him, named him Jacko, and very soon made his friendly acquaintance. And even then, and up to the present time, it has not been satisfactorily ascertained to what race the new discovery belongs. He is of the gorilla type, but not definitely enough to be declared a gorilla, which is, moreover, a creature unknown to the latitude of British Columbia—while there has been no menagerie there to introduce even a monkey. He is about 4 feet 7 inches in height, and weighs 127 pounds. His entire body, except his hands and feet, is covered with black, glossy hair about one inch in length, but his forearm is much longer than that of a man, and so strong that he will break a stick—by wrenching or twisting it—so large that no man could possibly accomplish this feat. He makes a noise, half bark and half growl, but is generally quiet. His favorite food is berries, and he drinks fresh milk with evident relish. His captor intends taking him to London for exhibition. Then his exact position in natural history will probably be discovered.

Contemporary news report of a Sasquatch capture in British Columbia, 1884.

The skull of alleged Caucasus manimal descendant Khwit.

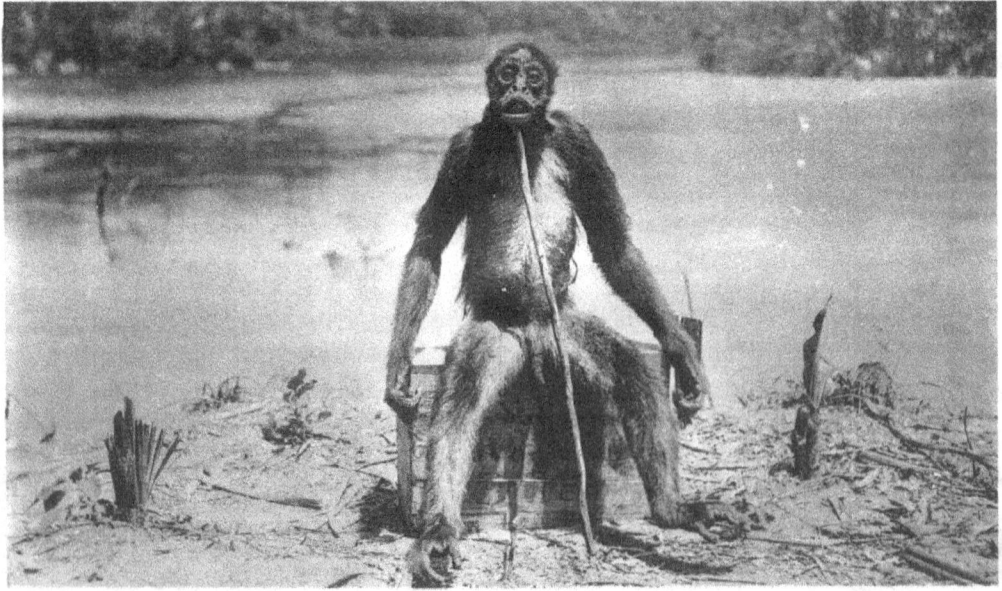

François de Loys's *Ameranthropoides loysi*.

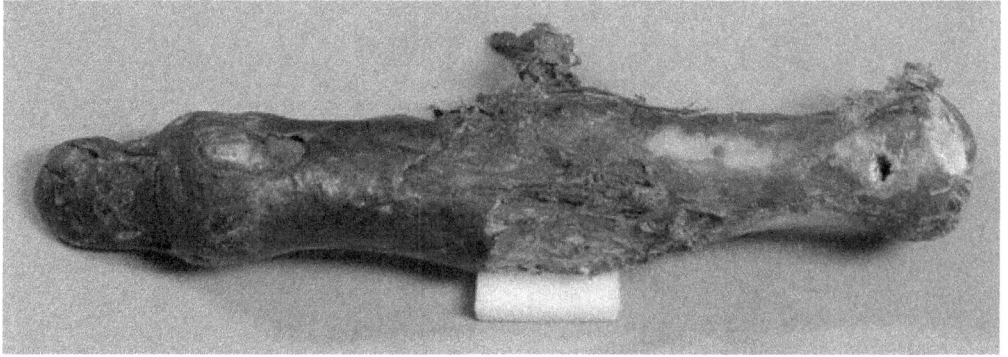

Alleged Yeti finger, purloined or purchased from Nepal's Pangboche monastery.

. Peter Byrne arranged for acquisition of the suspect finger.

James Stewart smuggled the finger from India into the West.

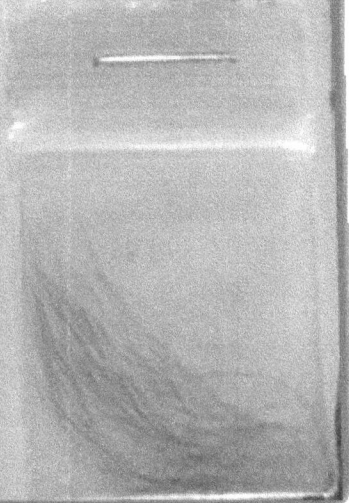

"Certified" Sasquatch hair, offered for sale on the Internet.

Doug Tarrant: lawman, Sasquatch witness and conspiracy theorist.

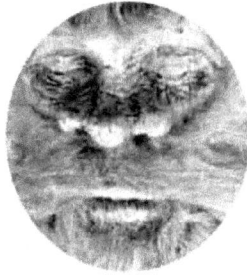

Chapter 8.
Found...and Lost

For every Sasquatch funeral reported, every "possible" gravesite discovered and then misplaced, we have multiple accounts of giant skeletons unearthed across North America and elsewhere. Students of the phenomenon posit various explanations for those anachronistic remains, ranging from ancient tribes of giant "Indians" to antediluvian "Nephilim"—but could some, at least, represent mortal remains of Sasquatch?

Robert Lindsay, for one, believes so. He writes: "Between 1858-present, over a 153 year period, 30 possible Bigfoot bones, skulls, skeletons or teeth have been found. So Bigfoot bones are found about once every 5.1 years in the modern era, about once every 5 years."[1] And in fact, as we shall see, his tally of giant remains is far too conservative.

One who doubted it was Dr. Grover Krantz, anthropologist and outspoken Sasquatch believer, who wrote:

> I have spent a good deal of time trying to track down the stories of enormous jaws and other bones, only to find that the specimen either was quite normal or else it could not be located. Other investigators with whom I have compared notes have had similar experiences. Any time one hears of such a specimen—something that looks like a human bone but is much too large—it is almost certainly a normal bone of a human or some other animal. I say "almost" because there is always a small chance that some day it will be the real thing. So I still chase down such stories no matter how slim the odds may seem.[2]

And let the chase begin.

On Other Shores
Reports of giant humanoid remains in Europe predate those claimed from North America by at least 325 years. The first such find allegedly occurred on the feast day of St. Peter and St. Paul—June 29—in 1535, at the Breitenwinner Cave on the Swiss/Bavarian border. There, as

reported by Berthold Buchner and attested by the treasurer of Areberg, a party of 25 men found "so many bones that the first of us had to pile them up in one place to make room for us to enter. The bones were very large as if from giants." After various mishaps, the party emerged safely, but with no evidence of their find.[3]

Leap forward 125 years, to the north of England, where we read of the next discovery.

> The banks of the Cor being worn away by some impetuous land-floods, there was discovered, about the year 1660, a skeleton, conjectured to be that of a man, of 9

Passage of another 59 years brings us to Salisbury, in the English county of Wilshire, where a brief newspaper item reads:

> GIANT SKELETON AT SALISBURY—A French paper on giants gives a list of several, whereof the biggest is one found near Salisbury, and the reference is to a French paper, 1719. Its length was 9 ft. 4 in. English, which is the largest human stature of which I ever heard. At Salisbury I remember in childhood a mound in a field, north of St. Edmund's Churchyard, called the "giant's grave." Is there any account of this skeleton and where it is kept?—as a skeleton of that size was surely worth preservation. E. L. G.[5]

Worthy indeed, but E. L. G.'s inquiry evoked no response. For the record, according to Guinness World Records, the tallest known human was Robert Pershing Wadlow of Illinois, who measured 8 feet 11.1 inches at his death in July 1940. Wadlow was also a certified "Bigfoot," wearing a size 37AA shoe (18.5 inches long) in adulthood.[6]

Sicily produced an even larger giant, 88 years after the Salisbury discovery. Author Joseph Comstock, writing in 1838, described the find at a depth of 170 feet in a sulfur mine.

> When they were attempting to remove a part of this wall, it fell into a hollow place or cell, upon two marble coffins, which contained the gigantic bones. The falling of the wall so deranged the place, that it could not be told whether it was erected for a place of sepulture, or whether it was a part of some building of another kind. And although one of the skeletons was much broken by the accident, very happily the other was entire, except the loss of a small part of one of the bones of the leg.

> Capt. Allen placed the bones of the most perfect skeleton in their proper position, and fond the skeleton to be eleven feet and four inches in length, Italian measure, which is equal to about ten and a half feet English! Capt. Allen descended to the bottom of this deep excavation, and carefully examined the hieroglyphics, which he says were engraved in the most curious manner, on the walls. The boxes or coffins were also ornamented with hieroglyphics.... Capt. Allen tells us that the head of the skeleton, including the skull and jaws, were about the dimensions of a two gallon pail or bucket. The diameter of the thigh bone he supposed to be about four English inches.[7]

We are clearly well beyond the realm of "normal" Sasquatchery with the Sicilian case, and shall be with others presented below, but I offer them here for what they may be worth, concerning the alleged existence of Sasquatch-sized remains.

New York's *Oswego Commercial Times* published the following item from France in August 1851:

> SKELETON OF A GIANT—Recently a gentleman in the neighborhood of the ancient city of Reims, in making an excavation for some purpose, discovered a human skeleton, well preserved, which was four metres (13 ft.) long.[8]

No further details were forthcoming, and while some may dismiss the account as standard "silly season" filler fabricated by the newspaper's staff, others might wonder if it could relate in some way to reports of giant "wild men" said to inhabit early Europe and the British Isles. Without the skeleton, of course, no one can say.

Giant discoveries in Europe would continue through the latter 1950s—possibly the subject of another book—but for the moment, let us turn back to the more familiar range of Sasquatch sightings in the Western Hemisphere.

Giants of North America
Closer to home, our first report of giant remains dates from the 1790s, belatedly reported by the *New York Times* in August 1880, quoting from Pennsylvania's *Harrisburg Telegraph.* That item read:

> The following was copied verbatim from a note made in his pocket almanac by the late Judge Atlee: "On the 24th of May, 1798, being at Hanover (York County, Penn.) in company with Chief-Justice McKean, Judge Bryan, Mr. Burd, and others, on our way to Franklin, and, taking a view of the town in company with Mr. McAlister, and several other respectable inhabitants, we went to Mr. Neese's tan-yard, where we were shown a place near the currying-house from whence (in digging to sink a tan-vat) some years ago were taken two skeletons of human bodies. They lay close beside each other, and measured about 11 feet 3 inches in length; the bones were entire, but on being taken up and exposed to the air they presently crumbled and fell to pieces. Mr. McAlister and some others mentioned that they and many others had seen them, and Mr. McAlister, who is a tall man, about 6 feet 4 inches high, mentioned that the principal bone of the leg of one of them, being placed by the side of his leg, reached from his ankle a considerable way up his thigh, pointing a small distance below the hip bone."[9]

Four decades later, another find was made at Moundsville, West Virginia, named for its enigmatic pre-Columbian burial mounds. White settlers began excavating said mounds in

1830, and eight years later reportedly found two "large" skeletons, male and female, the latter adorned with copper bracelets.[10]

According to researcher David Cain, West Virginia yielded another find less than 20 years later. He writes: "In the 1850s while excavating a root cellar in Palatine (East Fairmont), workers uncovered two very large human skeletons. Measuring the bones, people were amazed to find the entombed humans had been more than 7 feet tall. Many curious onlookers observed the skeletons which mysteriously disappeared overnight, apparently stolen for greedy purpose."[11]

Vanishing remains, as we shall see, are the bane of any researcher pursuing historical giants. We have a slightly more specific date for the next discovery—1852—though it seemingly avoided appearing in print for another 45 years. Pennsylvania's *Towanda Daily Review* ran the story in October 1897.

> Visitors go up on Pisgah almost every pleasant day. Mrs. Atwater, who is making some improvements, and Mr. Cottrell of Mansfield is there and drives the team down occasionally for the visitors. The Athens Historical Society sent word to the mountain they would visit it next Saturday. They claimed to know of an Indian grave on the top that they would open. The only Indian grave we ever knew of there says the Troy Gazette was one on the south point of the mountain on the farm of Chas. W. Hooker. A very large thigh bone of a human being was dug up 45 years ago at that point near a spring. It was of immense size and on its being shown to Dr. Theodore Wilder, he said it must have belonged to a man seven feet high. "There were giants in those days."[12]

Location of the site is hampered not only by passage of time, but by the presence within Pennsylvania of five summits named Pisgah, all presumably christened for a mountain of that name mentioned in the book of Deuteronomy.[13] (*Pisgah* means "peak" or "summit" in Hebrew.) No other source confirms discovery of the oversized thighbone.

Robert Lindsay injects a note of confusion with our next case, turning one find into two.

> 1856: Ohio or West Virginia. Possible Bigfoot skeleton found with bullet holes in its skull. Reported in the Bigfoot Track Record [*sic*].
> 1856: East Wheeling, West Virginia. A decayed 9'6" skeleton was found with three bullets in its head. No one knew what to make of it. Reported by Ray Crowe.[14]

In fact, they were one and the same, although Lindsay understates the skeleton's size, as reported in this item from the *New York Times,* published on November 21, 1856, with credit to the *Wheeling Times.*

> SKELETON OF A GIANT FOUND—A day or two since, some workmen engaged in subsoiling the grounds of Sheriff Wickham, at his vineyard in East Wheeling, came across a human skeleton. Although much decayed, there was little difficulty in identifying it, by placing the bones, which could not have belonged to other than a

human body, in their original position. The impression made by the skeleton in the earth, and the skeleton itself, were measured by the Sheriff and a brother in the craft *locale,* both of whom were prepared to swear that it was *ten feet nine inches in length.* Its jaws and teeth were almost as large as those of a horse. The bones are to be seen at the Sheriff's office.[15]

The IBS database adds a strange twist, reporting that "When three bullets were found in the skull it was dismissed as a hoax."[16] Years later, in his May 2011 blog, Ray Crowe speculated that the skeleton might represent "a Capone execution"—presumably referring to Prohibition-era gangster Al Capone of Chicago.[17] We can only hope that Crowe was joking, since Capone was born in 1899 and did not assume major gangland status until the 1920s.

The *New York Times* reported another giant find in December 1868, quoting an item from Minnesota's *Sauk Rapids Sentinel.* It read:

Day before yesterday, while the quarrymen employed by the Sauk Rapids Water Power Company were engaged in quarrying rock for the dam which is being erected across the Mississippi, at this place, they found embedded in the solid granite rock the remains of a human being of gigantic stature. About seven feet below the surface of the ground, and about three feet and a half below the upper stratum of rock, the remains were found imbedded in the sand, which evidently had been placed in the quadrangular grave which had been dug out of the solid rock to receive the last remains of this antideluvian [*sic*] giant. The grave was twelve feet in length, four feet wide, and about three feet in depth, and is today at least two feet below the present level of the river. The remains are completely petrified, and are of gigantic dimensions. The head is massive, measures thirty-one and one-half inches in circumference, but low in the *asfrontis,* and very flat on top. The Femur measures twenty-six and a quarter inches, and the Fibula twenty-five and a half, while the body is equally long in proportion. From the crown of the head to the sole of the foot, the length is ten feet nine and a half inches. The giant must have weighed at least 900 pounds when covered with a reasonable amount of flesh. The petrified remains, and there is nothing left but the naked bones, now weigh 304¼ pounds. The thumb and fingers of the left hand, and the left foot from the ankle to the toes are gone; but all the other parts are perfect. Over the sepulchre of the unknown dead was placed a large, flat limestone rock that remained perfectly separated from the surrounding granite rock.[18]

On August 23, 1871, Toronto's *Daily Telegraph* reported a startling discovery of Canadian giant remains. According to that item—

On Wednesday last, Rev. Nathaniel Wardell, Messers. Orin Wardell (of Toronto), and Daniel Fredenburg, were digging on the farm of the latter gentleman, which is on the banks of the Grand River, in the township of Cayuga. When they got to five or six feet below the surface, a strange sight met them. Piled in layers, one upon top of the other, some two hundred skeletons of human beings nearly perfect—around the neck of each one being a string of beads.

There were also deposited in this pit a number of axes and skimmers made of stone. In the jaws of several of the skeletons were large stone pipes—one of which Mr. O. Wardell took with him to Toronto a day or two after this Golgotha was unearthed....

The skulls and bones of the giants are fast disappearing, being taken away by curiosity hunters. It is the intention of Mr. Fredinburg to cover the pit up very soon. The pit is ghastly in the extreme. The farm is skirted on the north by the Grand River. The pit is close to the banks, but marks are there to show where the gold or silver treasure is supposed to be under. From the appearance of the skulls, it would seem that their possessors died a violent death, as many of them were broken and dented.[19]

Virginia soon produced the next discovery, as reported by the *New York Times* on September 8, 1871.

Not to be behind Canada, Virginia puts in a claim for a cave full of dead Indians, the Petersburg *Index* giving the tale as quoted below, on the authority of gentlemen whom it asserts to be of the highest character and credit, who have seen with their own eyes, and touched and tested with their own hands, the wonderful objects of which they make report as follows:

"The workmen engaged in opening a way for the projected railroad between Weldon and Garysburg struck Monday, about one mile from the former place, in a bank beside the river, a catacomb of skeletons, supposed to be those of Indians, of a remote age and a lost and forgotten race. The bodies exhumed were of strange and remarkable formation. The skulls were nearly an inch in thickness; the teeth were filed sharp, as are those of cannibals, the enamel perfectly preserved; the bones were of wonderful length and strength—the *femur* being as long as the leg of an ordinary man, the stature of the body being, probably, as great as eight or nine feet. Near their heads were sharp stone arrows, stone mortars, in which their corn was brayed, and the bowls of pipes, apparently of soft friable soap-stone. The teeth of the skeletons are said to be as large as those of horses. One of them has been brought to the city, and presented to an officer of the Petersburg Railroad. The bodies were found closely packed together, laid tier on tier as it seemed. There was no discernible ingress into or egress out of the mound."[20]

Who were the "giant Indians"? Did they exist at all? We shall explore that question more fully in Chapter 11, while discussing the views of conspiracy theorists. For now, we can only lament the apparent persistent failure to preserve such unique remains.

The *New York Times* weighs in again, in May 1882, reporting a discovery from Minnesota.

St. Paul, Minn., May 24—A skull of heroic size and singular formation has been discovered among the relics of the mound-builders in the Red River Valley. The mound was 60 feet in diameter and 12 feet high. Near the centre were found the bones of about a dozen men and women, mixed in with the bones of various animals. The skull in question was the only perfect one, and near it were found some abnormally large body bones. The man who bore it was evidently a giant. A thorough investigation of the mound and its contents will be made by the Historical Society.[21]

Eighteen months later, the *Times* carried more giant news, this time citing West Virginia's *Charleston Call.*

Prof. Norris, the ethnologist, who has been examining the mounds in this section of West Virginia for several months, the other day opened the big mound on Col. B. H. Smith's farm, six or eight miles below here. This is the largest mound in the valley and proved to be a rich store-house. The mound is 50 feet high, and they dug down to the bottom. It was evidently the burial place of a noted chief, who had been interred with unusual honors. At the bottom they found the bones of a human being, measuring 7 feet in length and 19 inches across the shoulders. He was lying flat, and at either side, lying at an angle of about 45 degrees, with their feet pointed at the chief, were other men, on one side two and on the other three. At the head of the chief lay another man, with his hands extended before him, and bearing two bracelets of copper. On each side of the chief's wrists were six copper bracelets, while a looking glass of mica lay at his shoulder and a gorget of copper rested on his breast. Four copper bracelets were under his head, with an arrow in the centre. A house 12 feet in diameter and 10 feet, high, with a ridge pole 1 foot in diameter, had been erected over them, and the whole covered by the dirt that formed the mound. Each of the men buried there had been inclosed in a bark coffin.[22]

Nothing there sounds much like Sasquatch, besides the fact that those remarkable remains have vanished like the rest, without a trace.

In May 1885 the *New York Times* reported a fresh discovery from Ohio.

Centersburg, Ohio, May 4—Licking County has been for years a favorite field for students of Indian history, there being two old forts and scores of mounds. Last week a small mound near Homer was opened by some schoolboys, who found a skeleton. To-day further research was made, and several feet below the surface of the earth in a large vault, with stone floor and earth covering, were found four huge skeletons, three being over seven feet in length and the other eight. The skeletons lay with their feet to the east on a bed of charcoal in which were numerous partially burned bones. About the neck of the largest skeleton were a lot of stone beads, evidently a necklace in life. The grave contained about 30 stone vessels and implements, the most striking being a curiously wrought pipe, the bowl having a series of carved figures upon it representing a contest between animals and birds. It is said to be the only engraved stone pipe ever found. A stone kettle holding about a gallon, in which was a residue of saline matter, bears evidence of much skill. Their bows, a number of arrows, stone hatchets, and a stone knife are among the implements. The knife is of peculiar shape, with a curved blade and wooden handle. Students of Indian archaeology claim it is the most valuable find ever made in that line.[23]

And yet, improbably, every scrap has vanished.

Robert Lindsay tweaks our interest next, but ultimately disappoints. His brief notation reads: "July 1885: New York. A large number of huge skeletons were found in a cave. They were up to 8 feet tall. Disposition unknown. Reported by the Daily Victoria Standard, Victoria, British Colombia, July 7, 1885."[24] Lindsay's link to a putative source was dead when checked in August 2014, and the cited article proved irretrievable at press time for *Sasquatch Down*. Multiple Internet blogs allude to discovery of giant remains in the mid-1800s, near Rutland and Rodman, New York, but none offer any further details.

Lindsay returns with another item, allegedly reported nine months after the find in New York: "April 1886: Etowah, Alabama. Giant skeletons were found after a flood, washing out of riverbanks. Disposition unknown. Reported by the Jacksonville Republican, Jacksonville, Alabama, April 21, 1886."[25] This time, Lindsay's link leads to a snippet from the *Gadsden News,* which reads:

> Mr. James F. Henry who discovered the bones of the big skeleton on his father's farm on the bank of the Coosa River near Gadsden, says that he could easily place his head in its skull and the bone was half an inch thick. The thigh bone was about twenty-two inches in length and three times as large as the bone in an ordinary man. The bone from the shoulder to elbow measured about twenty inches; and when all the bones were placed in their proper places they showed that the owner, when alive, must have been at least twelve feet from the top of his head to the bottom of his feet. Two or three of these enormous skeletons were found.[26]

New Jersey produced a crop of giant bones in February 1890, as reported by the *New York Times.*

> Mays Landing, N.J., Feb. 8—For over a week past crowds have been flocking to the site of the unearthed Indian graveyard near Edgewater avenue in Pleasantville. The first lot of skeletons unearthed was about one thousand yards from the city Post Office and embraced eight bodies, closely laid together in a deep chamber, snugly packed in with tortoise, oyster, and clam shells....

> Prof. C. H. Farrell of Baltimore, Charles K. Simpson of New York, John H. Cooley, Jr., of New Haven, Conn., and several gentlemen from the University of Pennsylvania immediately went to the scene. Messrs. Risley and Farr, the owners of the land, gave to the Archaeological Association of the University of Pennsylvania the right to search for relics on their land. These researches have been watched by thousands of people with great interest. Besides weapons of war, savage ornamental war decorations and numerous valuable shells, stones, &c, over fifty skeletons have been exhumed.

> Dr. Charles R. Abbott, curator of the association, is continuing the search and the skeletons are to be shipped to the university at once. They run in size from a small child to several seven feet in height, and one, supposed to be an old medicine man,

Wauneck, must have been at least eight feet in height.[28]

Where those priceless relics wound up is anyone's guess.

South of the Border

Mexico has produced sightings of large hairy bipeds known as *Quinametzin*, and it has a record of giant remains to match. In August 1895 the *Coconino Sun*, published in Flagstaff, Arizona, announced one such discovery in a snippet reading: "In a mound near San Juan, Mex., there has just been unearthed the skeleton of a giant man, who evidently belonged to a prehistoric race. The length of the skeleton is 12 feet 7 inches."[29]

In May 1908, two U.S. mining executives reported their discovery of a Mexican cave—location unspecified—containing "some 200 skeletons of men each above eight feet in height," supposedly "a race of giants who antedated the Aztecs." Charles Clapp, primary spokesman for the miners, had "arranged the bones of one of these skeletons and found the total length to be 8 feet 11 inches. The femur reached up to his thigh, and the molars were big enough to crack a cocoanut. The head measured eighteen inches from front to back."[30]

Miners scored another find at Sisoguichi, high in the Sierra Tarahumara mountains of Chihuahua, in June 1925. Details are sparse, but newspaper reports claimed that "some of the individuals" stood 10 to 12 feet tall in life. Their disarticulated foot bones, when reassembled, averaged 18 to 20 inches long.[31]

Eleven months later, another newspaper story described a discovery of giant skeletons in Nayarit, a state on Mexico's Pacific coast, between Durango and Jalisco. According to that item, Captains F. W. Devalda and D. W. Page made the find, reporting skeletons that "averaged over ten feet in height."[32]

Another quarter-century passed before the *Kentucky New Era* headlined discovery of "huge blond Mexican Giants" at Barranca de Cobre ("Copper Canyon"), in Chihuahua's Sierra Madre Occidental range. Herpetologist Paxson Hayes and guide Rafel Garcia found the remains in a cave while searching for snakes. The find included nine mummified bodies and 34 skeletons, each within a wicker basket and "wrapped like a silkworm." According to the article, "When the bones of the mummies were laid out properly the various bodies measured from 7 feet six inches in most cases up to the largest skeleton which is a full eight feet!"[33]

And we are left to ask: where are they now?

Dem Dry Bones

In December 1897 the *New York Times* proclaimed a remarkable discovery at Maple Creek, Wisconsin. According to that article: "One of the three recently discovered mounds in this town has been opened. In it was found the skeleton of a man of gigantic size. The bones measured from head to foot over nine feet and were in a fair state of preservation. The head was as large as a half bushel measure. Some finely tempered rods of copper and other relics were lying near the bones."[34]

British Columbia yielded its first recorded giants in November 1900, discovered by miner James Perkinson near Atlin. That article tells us that—

> Five skeletons, nearly complete, were exhumed and each is the set of bones that belonged to a giant of prehistoric times. One of the skeletons measures over seven feet in length, so that the man must have been considerably over that height. Then there were two others of within an inch of seven feet and the remaining two "were more than six feet in length and the men were of gigantic frame."
>
> ... The bones of the fingers and toes had crumbled away, but the finger of one skeleton hand was sufficiently strong to hold a ring of what appears to be lead or some similar base metal. The skeletons were unusually well formed, but one unique feature was that the arms were several inches shorter than ordinarily appears, while the size of the bones of the forearm was enormous in comparison to the usual models.[35]

In February 1902 the *New York Times* proclaimed "the discovery of a race of giants" in Guadalupe County, New Mexico, near Mesa Rico. "Skeletons of enormous size" had been unearthed on a ranch owned by Luiciana Quintana, including one "that could not have been less than 12 feet in length." That individual's forearm measured four feet long, his ribcage was seven feet in circumference, and his teeth "ranged from the size of a hickory nut to that of the largest walnut in size." Over a period of years, Quintana had found "perhaps thousands of skeletons" buried on his property.[36]

Laborers on the New York and Harlem Railroad made the next find, uncovering several "skeletons of unusual size" and various artifacts near Katonah, New York, in September 1904. No specific measurements were published, and the remains were reburied at Katonah in unmarked graves.[37]

May of 1912 brought news of another find near Delavan Lake, Wisconsin. Eighteen skeletons were found inside a burial mound, and while newspaper coverage mentioned no overall measurements, the skulls are of interest to our inquiry.

> The heads, presumably those of men, are much larger than the heads of any race which inhabit America to-day. From directly over the eye sockets, the head slopes straight back and the nasal bones protrude far above the cheek bones. The jaw bones are long and pointed, bearing a minute resemblance to the head of the monkey. The teeth in the front of the jaw are regular molars. There were also found in the mounds the skeletons, presumably of women, which had smaller heads, but were similar in facial characteristics.[38]

In July 1916 the *New York Times* reported yet another giants' mass grave, this one discovered at Tioga Point, near Sayre, Pennsylvania. According to the article, the grave contained "the bones of 68 men which are believed to have been buried 700 years ago. The average height of these men was seven feet, while many were much taller...On some of the skulls, two inches above the perfectly formed forehead, were protuberances of bone." One skull and "a few bones" had reportedly been shipped to "the American Indian Museum," presumably a reference to the part of the Smithsonian Institution known since 1989 as the National Museum of the American Indian.[39] (For more on the Smithsonian and vanishing remains, see Chapter 11.)

Robert Lindsay, having missed all of the foregoing cases, joins us now with a report from 1920s California: "1923: Santa Barbara. J. P. Harrington found and examined "Indian skulls" with very peculiar qualities. He felt that they resembled Neandertal Man. He concluded that they were modern Indians from the Santa Barbara region. Reported by Anonymous, Nature, 112:699, 1923."[40] A newspaper report from April 1923 elaborates.

> All doubt as to the greater age of the skulls of the "Santa Barbara man" uncovered here this week, as compared with the Neanderthal man of Central Europe, has been dispelled in the minds of scientist doing excavation work on the Burton Mound fronting the Santa Barbara ocean beach according to J. P. Harrington of the Smithsonian Institution in a formal statement tonight.
>
> Dr. Harrington, who has been in charge of southern California archaeological work for the Smithsonian Institution for several months, is certain that a new link in the Anthropological chain has been established definitely by the excavations of the last few days. Further examination of the gorilla-like skulls unearthed on Burton Mound, he asserts, has definitely proven that the Santa Barbara man existed in a period far earlier than the era of Neanderthal man. Not only that, but he possessed a culture which far exceeded that of the Neanderthal.
>
> The thickness of the skulls is twice as great as those of Indians found in the burial grounds, known to be 1,000 years old or more. The average thickness of each skull is approximately three-quarters of an inch.
>
> Dr. Jesse Walter Fewkes, chief curator of the Smithsonian Institution telegraphed Dr. Harrington today for a complete report of the discoveries made here. An authoritative and official statement has been dispatched to him.[41]

In October 1934 the *Sarasota Herald* reported discovery of a "redskin burial ground" outside Simcoe, Ontario, that yielded an unspecified number of "skeletons eight feet long."[42] The discovery also rated mention in the *Milwaukee Sentinel.*[43]

Late in 1939, laborers employed by the New Deal's Works Progress Administration began excavation of the so-called Moorhis Mound, near the Guadalupe River in Victoria County, Texas. What they found was reported by the *San Antonio Express* on January 7, 1940.

> That Texas "had a giant in the beach" in the long ago appears probable from the large skull recently unearthed on a mound in Victoria County, believed to be the largest human skull ever found in the United States and probably in the world. Twice the size of the skull of a normal man, the fragments were dug up by W. Duffen, archeologist who is excavating the mound in Victoria County under a WPA

project sponsored by the University of Texas. In the same mound and at the same level, a normal sized skull was found. The pieces taken from the mound were reconstructed in the WPA laboratory under the supervision of physical anthropologists. A study is being made to determine whether the huge skull was that of a man belonging to a tribe of extraordinary large men, or whether the skull was that of an abnormal member of a tribe, a case of gigantism. Several large human bones have been unearthed at the site. Marcus B. Goldstein, physical anthropologist, employed on the WPA project, formerly was an aide of Alen Horliken, curator of the National Museum of Physical Anthropology. The finds made through excavations in Texas are beginning to give weight to the theory that man lived in Texas 40,000 years ago, it is said.[44]

Our focus shifts back to British Columbia in May 1943, with the following report from Nanaimo.

Discovery of large-sized Indian bones by victory gardeners at Departure Bay, three miles north of here, is believed to give support to the legend that a 'giant tribe' inhabited Vancouver Island 300 years ago. The lower jaw, part of a skull and the shin bone of an Indian were unearthed, and preliminary examination suggests that their owner may have been around seven feet tall, weighed more than 400 pounds and was between 70 and 80 years old when they died. It is believed he may have died in battle, as the legend persists that a giant tribe exterminated the Nanaimo Indians 300 years ago at Departure Bay. Bones found were covered with small rocks three feet under the surface.[45]

Robert Lindsay's next report comes from Alaska, in the postwar era: "1948: Bartholomew Creek, Smeaton Bay, East Behm Canal, Alaska. In Misty Fiords National Park, two men reported finding a jawbone larger than a man's. Possible Bigfoot jawbone. Reported in the Bigfoot Track Record [sic]."[46] The men, alas, remain anonymous, and the jawbone remains invisible.

Lindsay's next case is equally—one might say typically—frustrating: "After 1960: Pendicton [sic], British Colombia. In south-central British Colombia, two fishermen found a dead Bigfoot along the trail. They first smelled the corpse, then found the body. Upon investigating, they heard sounds in the brush. Fearing it was another Bigfoot, they quickly left the scene. They went back with wildlife officials 10 days later and there was only a dark spot on the trail. Had a bear eaten it? Had other Bigfoots hauled it off? Reported by Peter Byrne."[47]

Lindsay logs two discoveries of "possible" Sasquatch remains in 1965, the first reported from the Bluegrass State: "1965: Wolfe, Kentucky. Kennith White found a nine foot skeleton with long arms and a huge head while digging along a creek bank. It was later reburied. Reported by the Kentucky Bigfoot website."[48] Tracing his source, we find the following account from Holly Creek, reported by Paul Henson.

Bigfoot bones or something else entirely? While digging under a large rock along the creek bank, a local named Kennith White came across the bones of a 9' tall being with extremely long arms, large hands and a skull 30" in circ. Even stranger was the fact that the eye and nose sockets were slits—not cavities as found in humans. Moreover, the area where the lower jaw-bone normally hinges to the skull was solid bone which tells us this strange creature had no possible way of opening its mouth. A mysterious white, powdery substance was covering the skeletal remains when found. The farmer later reburied the bones at another location.[49]

Lindsay's second report from 1965 reads: "1965: Minarets Region of the Sierra Nevada, California. A partial Bigfoot skull (calvarium) was found by a physician. A pathologist said it was not human. It was sent to UCLA, where anthropologists said it was an old Indian skull, since the only ancient hominids residing in the Sierras were Indians, so it must be an Indian. They did say that it had odd features such as a nuchal crest. It's presently lost in storage. Reported by the BFRO."[50] Matt Moneymaker's BFRO report contains more detailed information, including the month of the discovery (August), name of the doctor who found the partial skull (Robert W. Denton), the pathologist who first examined it (Dr. Gerald K. Ridge, at Ventura County General Hospital), and the UCLA anthropology professors who received it next (Dr. Herman Bleibtreul and Dr. Jack Prost).[51] Details of the episode are both intriguing and, perhaps, instructive.

According to Moneymaker, Dr. Denton found the calvaria—or skullcap, including the superior portions of the frontal, occipital, and parietal bones—at Hemlock Crossing, on the North Fork of the San Joaquin River, where a farm worker's mule, mired in mud, kicked the relic free. Dr. Ridge deemed the specimen a relic from to "some anthropoid species other than human," based on its size and apelike form. Specifically, he called the partial skull "a rather interesting specimen largely by virtue of the unusual length of the skull as well as a very unusual development of the nuchal ridge in the occipital zone. This latter fact for a time had me thinking this must be the skull of some anthropoid species other than human, inasmuch as this amount of nuchal ridge development had not been observed by me." Doctors Bleibtreul and Prost "had never before seen anything quite like it," but they finally pronounced it the skull of a "young, ancient Indian male." In December 1965, Dr. Denton sent Ridge a map marked with the location of his find, but he heard no more about it.[52]

In August 1973, Sasquatch researcher Alan Berry met Denton and began his own investigation of the mystery, hoping to find the skull fragment. Dr. Bleibtreul, transferred by then to the University of Arizona, denied any memory of the relic until Berry produced Dr. Ridge's receipt for its delivery to UCLA, then he vaguely recalled it. A student technician in UCLA's anthropology department claimed to have searched for the bone and found nothing, but Moneymaker questions whether he performed a thorough search—or if he even looked at all. The student suggested that Dr. Prost might have taken the calvaria with him when he left UCLA, a notion firmly rejected by Prost, who also denied any memory of the events in question. Speculation persists that the "Minaret skull" may still be gathering dust at UCLA, either concealed deliberately or simply filed away and forgotten in storage.[53]

One of our most intriguing—or bizarre—reports of a dead Sasquatch discovery comes to us from Okanogan County, Washington, via a terse report from the IBS sightings database.

According to that snippet, lacking the "full report" promised parenthetically, sometime during 1968 hiker Mike Stevenson and unnamed companions saw "2 BF FIGHTING; BROWN 11-12 FT, AND REDDISH 7 FT KILLED, STOMACH RIPPED."[54] If true, that incident could have provided all the evidence required to prove Sasquatch exists—but, alas, no one pursued it.

John Green, writing in 1978, is the source of our next report. According to him, "In June, 1971, the Salem, Oregon, *Capital Journal* carried a story about a local girl and her friend having seen a dead Bigfoot near Happy Camp, California, in 1967. I didn't learn of the story until several years later, and didn't pay much attention to it...Still, I never felt quite right about leaving such a story hanging, and I tried a few times to find the girl, or to get someone else to do it. Late in 1976 I finally took the matter on in earnest and got lucky."[55]

Green met one witness, married with children in Tacoma, Washington, and spoke to the other by phone in California. They claimed the dead Sasquatch was lying on its back, legs splayed, and recalled trying in vain to turn it over with a stick. There was "a great deal of hair, like a horse's mane, except that there was too much of it and it was spread over too wide an area." The corpse was "all rotted away," face gone and white ribs showing, but one witness clearly remembered a broad thumbnail on one hand, distinct from an animal's claw. Both recalled feeling they had been followed that day, by an unseen stalker that frightened their dog.[56]

After hearing their tales, Green wrote, "I am entirely satisfied that there was indeed what remained of some big-boned creature on or by that road, and I am pretty sure that at least the second girl, who knew the area, should be able to take someone pretty close to the spot. Even 10 years afterward I think there is an excellent chance, in fact all but a certainty, that some bone or tooth from a creature that big will have survived."[57] Apparently, no one saw fit to go and see. Ray Crowe, writing in 2011, seemed more skeptical of the story. Considering the body's well-advanced decomposition, he wrote, "Peee-yuu on that, why would they get so close?"[58]

Robert Lindsay logs our next report: "Before 1972: Shuswap Lake, British Colombia. In the Colombian Range of the Rocky Mountains, a Bigfoot skeleton was found washing out of a riverbed. The teeth and jaw were huge, and the skeleton was 8 feet long. It was sent to Wrexham Museum in Wales. It seems to have vanished into thin air and has never been found. Reported by Ivan Sanderson."[59]

Tracing Lindsay's report back to Sanderson, writing in 1961, we find that the discovery was actually made in 1912, by one Ernest A. Edwards, and reported to a friend in 1941. Sanderson wrote to the museum and received the following reply from Clifford Harris, its curator at the time: "With regard to your query, I have checked the Minutes of this establishment for the years 1912, 1913, and 1914, and there is no mention of the receipt of a skeleton."[60]

Foiled again.

Next up, Lindsay offers another vaguely dated incident: "After 1972: Antelope Flat, Oregon. On the Ochoco National Forest west of Bend, a Bigfoot skull was found. It was taken to

Portland College. They returned [it] after a bit with a long report, but only after it had been taken apart into its constituent pieces. They would not commit on what it was. Reported by Vic McDaniel."[61] The IBS database elaborates:

> Vic McDaniel reported that his uncle had found a Bigfoot skull near Antelope Flat in Lake County, OR. They took it to a Portland College where it was returned, but only after it had been taken completely apart into its component pieces. He was indeed miffed, but both Vic and his friend Roy said they gave uncle a long written report.[62]

Portland presently hosts no less than 18 colleges and universities. As *Sasquatch Down* went to press, both the skull fragments and the unnamed institution's "long report" had apparently vanished.

Robert Lindsay's next account deserves quotation in full, if only for the problems it creates.

> 1987: Estacada, Oregon. A man, Grover Kiggens, found his dog playing with a strange object. Upon examination, it was a human-like skull with strange features. It still had some skin and hair on it. The man felt that the creature had been 4-5 years old when it died. It seems to have been the skull of a young Bigfoot. There had been a lot of strange screeching in the forest for several nights previous to the discovery of the skull. The man thought it was human, so he sent it to the crime lab. The crime lab sent it back, saying it was not human. Then he sent it to the Regional Primate Center, but they refused to comment, simply returning the skull with a note. Next it went to the University of British Colombia, but they kept it for two years and could not decide what it was. Next it went to the University of California, Berkeley. After some time, the finder received a note from a Dr. Turner of Berkeley: "...please tell him he can be proud...is ultimately responsible for discovery of a new species and its legal protection. Slow going partly because legal protection requires species known to science, hence named and described based upon physical material. Several others and I cautious about going out on limb...process of elimination was very tedious, but skull is 'new.'" Berkeley is still on possession of the skull, but it seems to be lost. Reported by Cliff Olsen.[63]

This find generated so much interest that the IBS database listed it twice, under separate case numbers. One omits a date, while the other claims the skull was found in 1978, suggesting that Lindsay's date is a typographical error. The longer version quotes researcher Cliff Olson as saying:

> The lower jaw was still partly there. There had been a lot of "screeching" in the forest on several of the previous nights. Thinking it human, they sent it to the crime lab, and their report...the skull was not human! Next the human-looking skull with no sloping forehead and normal looking teeth, no long canines, went to the Regional primate center, but they wouldn't talk to her...just returned the skull with a note. They sent it to the University of British Columbia, and finally had to go retrieve it themselves after two years, as UBC drug their feet and wouldn't return it, they wouldn't commit themselves as to what it was. The Regional Primate center

was better, but said they couldn't identify it. They did comment that the sutures on the skull resembled those of a giraffe (tho' they knew it wasn't). Next it went to the Univ. of Calif., and after a binge of excitement about a possible new primate, they told her it was from an elk (even they could see it wasn't), and [it] has disappeared into the maws of that institution. There was some correspondence at the time that she'll try to find and send us, maybe we can track something down. I'll look into it if I can, but Jim Hewkin recalls trying years before to track it down unsuccessfully. Millie says she took some photos also...thinks she sent them to John Green, but not sure.[64]

The shorter version quotes Green, contradicting Olson: "As to the skull Grover Kiggins' dog brought home, it was a very partial skull, most of it chewed away. I took it to the team of primate anatomists who were then working at the Primate Center at the University of Washington and they found it pretty interesting but eventually identified it as part of the skull of a small sheep. I don't recall any teeth and there was certainly no part of the lower jaw that I saw."[65]

Vanishing evidence is the norm, where reputed Sasquatch remains are concerned. Another case in point comes from *National Geographic* magazine, whose September 1979 issue briefly reported the discovery of a unique callote—the thickest upper portion of a skull—at Lagoa Santa, Brazil. Dr. Alan L. Bryan, founder of the University of Alberta's Department of Anthropology, made the find, declaring that he believed the skull's former beetle-browed owner stood "a few rungs down the evolutionary ladder" from *Homo sapiens*. "I think," he wrote, "there is little question that the calotte is a transitional (i.e. early or archaic) Homo sapiens, somewhat like neandertal or Rhodesian Man, but more closely related to a North Chinese skull (Jinniushan) guesstimated to be less than 200,000 years old." Ray Crowe writes, "I'd like to call it a possible fossil Bigfoot callote," but the relic's disappearance and Dr. Bryan's death in May 2010 render the question moot.[66]

Robert Lindsay's next report is typically brief and vague: "Oregon: After 1980. A man and his son found a dead Bigfoot lying in a stream. They heard what sounded like another Bigfoot nearby watching over the dead one, and they quickly left the area. Reported by Cynthia Stayte."[67] No further information is available, beyond the fact that Stayte, a 49-year-old Missouri resident, began researching Sasquatch after she and her husband saw one of the creatures in July 1997.[68]

And again, from Lindsay: "Fall 1999: Connell Creek, Revillagigedo Island, Alaska. Near Ketchikan, Alaska, two men found an 8-inch hairy foot in sand by a creek. They threw it back in the creek. They said it belonged to neither a man nor a bear. Possible Bigfoot foot. Reported in the Bigfoot Track Record [*sic*]."[69] Hardly a *big* foot, but intriguing nonetheless. The dearth of names renders further research problematic, although the IBS database does list the incident as occurring at Latitude +055° 26' North, Longitude 131° 40' West.[70] For what it may be worth, the actual coordinates for Revillagigedo Island are Latitude 55° 38' 3″ North, 131° 17' 51″ West.[71]

Our last two reports with approximate dates come from Robert Lindsay, who writes:

2002: Scotts Valley, California. In the Santa Cruz Mountains, a man digging in a sand hill for shark teeth found a huge apparent Bigfoot tooth. He showed it to a several dentists, who all said it was human, but that it was too big to be human. It is presently part of Dr. Melba Ketchum's Bigfoot DNA project, but it has not yet been tested. Reported by Mike Rugg.

2008: Oregon. As part of the Ketchum DNA project to prove the existence of Bigfoot by sequencing their DNA, a purported Bigfoot bone, a femur, was used. The bone was found in a stream in Oregon. However, for whatever reason, the bone was not used in the study.[72]

Readers of *Sasquatch Down* are probably familiar with the controversial DNA report issued by Texas veterinarian Melba Ketchum and her Sasquatch Genome Project in 2013. After a five-year study, including 111 specimens of alleged Sasquatch hair, blood, skin and other tissues collected from 14 U.S. states and two Canadian provinces, Ketchum announced her discovery of DNA representing "a hybrid mix between *Homo sapiens* and an unknown primate." The ink was barely dry on that report when *Houston Chronicle* reporter Eric Berger sent some of Ketchum's samples to an independent geneticist for analysis. That expert's verdict: "a mix of opossum and other species. No find of the century."[73]

Without rehashing that brouhaha *ad nauseam,* we might ask why the aforementioned tooth and femur were excluded from Ketchum's study, but no one seems inclined to answer—or to say where the relics may be found today.

Cases Out of Time
As in preceding chapters, some of the reports on file come to us without any dates attached. Robert Lindsay leads off with: "Unknown date, modern era: Near Roseburg, Oregon. A hunter found a dead Bigfoot by a stream. He poked it and got no response. He tried to carry it out with the help of his hunting companions, but it was too heavy at 700-900 pounds. They went back to town to get a truck to cart it out with, but when they came back, it was gone. They found the footprints of another Bigfoot, which had apparently carried it away. Reported by Ray Crowe."[74] Crowe writes that the carcass "was not dragged[,] it was carried away."[75] Oddly, the IBS database for Douglas County incidents includes no reference to this case.[76]

Next up from Lindsay, we have this item: "Unknown date, modern era: Near Great Falls, Montana. A man's dog brought in a huge leg bone, from the pelvis to a foot. The man suspected it was a Bigfoot, so he gave it to a local university, but they could not identify it, and they never gave it back. The dog brought in another possible Bigfoot body part, but it smelled so bad that the man buried it. Coyotes then dug it up and ate it. Great Falls Tribune, Great Falls, Montana, 'The Beast's Foot.' Date unknown. Reported by Ray Crowe."[77] Great Falls lies in Montana's Cascade County. The IBS database lists two events from Cascade County—or, rather, the same event listed under two different case numbers—but includes no reference to the bone in question.[78] In response to my inquiry, staffers at the *Great Falls Tribune* found no relevant articles in the newspaper's files.[79]

Robert Lindsay's next report also comes from Big Sky Country. He writes: "Unknown date, modern era: Electric Peak, Gardiner, Montana. Two boys cross-country skiing on this mountain just north of Yellowstone National Park found the decomposed body of a Bigfoot. It was partly covered by a rock avalanche. The pelvis was crushed, and the skull was missing and had apparently been taken by headhunters. Reported by Ray Crowe."[80] The IBS database does include this case, crediting the information to researchers "Ben and Lisa M." Their account reads (uncorrected):

> Here's an interesting one, but can't seem to confirm. Will add a bit of the speculation. "I guess some boys on ski's found one (dead Bigfoot) near Gardner Montana," my sister-in-law said. "It was pretty well decomposed and partially covered by a rock avalanche. The pelvis was crushed and the skull missing. Still over 7 ft in length. The Livingston Police Dept. has taken over the area, near Electric peak and won't let anyone in. Apparently she heard two truckers talking at the Flying J. Truckstop in Billings where she works as a waitress." Couldn't find anything in the newspapers around Electric Peak, Montana about the seven-foot plus body found there. But, I did send an e-mail off to the Livingston Police Department for any information they may be able to supply concerning the rumor. Chances are they won't answer, but you never know.... There is something being covered up by the local populace. I spoke via e-mail to a former employee of (Outlaws Pizza, Gardiner, MT) High fences have been erected around part of the area. We are talking about a group of Religious Extremists. The fence, very large with the curly wire at the top. My guess is they are funded.... In regards to the story about some sort of gigantic, decomposed corpse found near Electric Peak near Gardner, Montana, I telephoned the Livingston Press [401 S. Main St., Livingston, MT 59047] this morning at their toll-free number: 1-800-345-8412. My call was transferred from a receptionist to someone in the newsroom where I explained that I had sought to verify a report that had appeared on the Internet. This person in the newsroom did not know anything about a recent story of a corpse found near Electric Peak and didn't know of anything involving the Electric Peak area being closed. He did add that back in the 50s some "guy had died in the area" and pieces of his body had been found here and there over the years.[81]

Our next undated report, from Ray Crowe, is a slightly altered version of the incident mentioned earlier in this chapter, from 1968. It reads: "Mike said two army buddies hiking in the Copper Mountain area near Brewster, Washington, had witnessed a Bigfoot battle. There was wood breaking, screams, roars and they saw a twelve-foot Bigfoot and a smaller brown one rolling in the dirt, punching, clawing, and throwing dirt and logs. The little one tripped, and the big one jumped on top and pounded his head with a large rock—killing him. The big one screamed, then tore at the little ones middle, pulling out a handful of guts that he greedily ate. An elaborate story or are there two types of Bigfoot?"[82] Readers will recall that in the earlier version, Mike Stevenson and an unnamed companion were listed as the actual witnesses.

Crowe provides another riddle in his March 2011 blog, writing: "At Pybus Bay near Ketchikan, Alaska, a man found two dried out corpses, two to three feet tall with big jaws and big canine teeth, and with some skin and hair still attached to the bones. They had been under the snow. Called baby Koostikay's, he would not give them to Fish and Game investigators."[83] No corresponding case appears in the IBS database online, despite one allusion to local aborigines fearing contact with mysterious "dwarfs of Pybus Bay."[84]

Robert Lindsay rejoins us with the following report: "Unknown date: Coshocton County, Oklahoma. Possible Bigfoot arm found in the woods. All that remained was skin, bones and hair. A specialist ran tests on it and said it did not come from any known animal in the area, nor from a human or an ape. He felt it was the best evidence yet for Bigfoot. Present status of the remains is unknown. Reported by Mary Green."[85] Green filed her report with the GCBRO. It reads (uncorrected):

> My mom's cousin was searching for mushrooms when he found the remains of an arm. The arm resembled a human arm except it was very large and the skin was covered in thick dark hair. The fingers also had claws growing out of the fingers. All that was left was the skin and hair and the bones. He called a specialists to come examine the arm and after examining it and running test, he concluded the hair and bones could not have came from an ape, human, bear, or any other creature he knew. He also said it may very well be the best evidence of bigfoot ever found. I'm not sure what happened after this but from what I have heard, days later more specialists came to see the arm and said they were going to take it to do more tests on it. It has not been seen since.[86]

Our next two reports are probably confused accounts of a single rumored incident, reported as separate events both by Robert Lindsay and Ray Crowe. Lindsay's accounts are more detailed, despite his difficulty in spelling "British Columbia."

> Date unknown, modern era: Toba River, British Colombia. In far southwest coastal British Colombia, a couple working a trap line found a Bigfoot skeleton washing out of a riverbank. The bones were too heavy to carry, but the wife packed out the huge jawbone against the advice of her husband. The University of British Colombia and the British Colombia Museum were called, and the couple reported that they had a Bigfoot jaw from a Bigfoot skeleton. The university and museum both said that there is no such thing as Bigfoot, so they didn't want to investigate. They kept the jawbone in their cabin, and 10 years later it burned down, taking the jaw in the process. Reported by John Green.

> Date unknown, modern era: British Colombia. The British Colombia Museum is said to be in possession of a huge jawbone, possibly of a Bigfoot, but they can't locate it, as it's crated somewhere in storage. Reported by John Green.[87]

Why the Royal British Columbia Museum would reject one jawbone and store another is anyone's guess. As for John Green's alleged reports, we can only say that his epic work on Sasquatch contains no reference to the Toba River or to disappearing jawbones.

Ray Crowe credits our next muddled reports to the late Alan Landsburg, a prolific TV writer-producer best known for his series *In Search Of...* (1977-82) and *That's Incredible* (1980-84).

Crowe's first version, appearing in a blog no longer found online, read: "They [Sasquatch bones], once again, had been reported washing out of a riverbank on Shushwap Island [*sic*], B.C., the teeth of huge size without cavities, and the entire skeleton eight foot long with skull (Alan Landsburg, 'In search Of Myths And Monsters,' saw in a college museum in Dublin, Ireland, a human skeleton of 8'6")".[88] When that blog disappeared, another took its place, including this: "Alan Landsburg said that a huge skeleton was washed out of a riverbank on Shushwap Island, British Columbia. The eight foot skeleton had huge teeth in the skull. Alan also said he saw at a college museum in Dublin, Ireland, an 8'6" skeleton."[89]

British Columbia has no "Shushwap Island." There is a *Shuswap* Lake, but its only island is named Copper Island.[90] The Landsburg book cited by Crowe contains no mention of Sasquatch teeth found in British Columbia, but it does mention the author's viewing of an eight-foot-six-inch *human* skeleton at an unnamed museum in Dublin.[91] Crowe seems to be conflating Landsburg's published observation with Ivan Sanderson's earlier tale of remains found at Shuswap Lake and sent to Wales.

Crowe's next report is a mere recitation of rumor: "Also it is said that uranium millionaire [Charles] Steen had one of the giant skulls that he used as an ashtray."[92] Such tales are tantalizing, but add nothing to our knowledge. Steen, who died from Alzheimer's disease in 2006, on his 87th birthday, is unavailable for comment.

Robert Lindsay's next item reads: "Unknown date: Glacier, Montana. Just east of Glacier National Park, a Bigfoot skeleton was said to be kept in a sacred Native American burial area, possibly buried along with the Indians. Reported in the Bigfoot Track Record [*sic*]."[93] Tracing this tale back to the IBS database, we find one Scott Sebring listed both as researcher and witness. Ray Crowe describes Sebring as "blonde, but says he is half Blackfoot."[94] Sebring's story (uncorrected below) does not precisely correspond to Lindsay's.

> While visiting his great-grandfathers remains at the Blackfoot Reservation in Montana, Scott learned of an upcoming pow-wow dedicated to the "hairy man." Scott says, once every one, two, or three years, it is said that, "when the buffalo, the deer, and the coyote come together, then the hairy man appears." ... During the pow-wow dance, a man dressed in a hair-suit dances out and entertains the young ones...and the elders also... The Blackfeet think that the "Bigfeet" are just another "type of people," and want to just leave them alone. Scott says his relative is interred in a lava tube/cave that has been enlarged. One of the other things he says he saw in the cave were the bones of a very large man, 8½ feet tall, that had a strange, unhuman looking skull. The burial area and pow-wow are open only to Native Americans.[95]

Ray Crowe presented our next case in March 2011, echoed and expanded by Robert Lindsay two months later. We shall discuss it more fully in Chapter 11, but for now, Lindsay's version should suffice to complete our record of alleged Sasquatch skeleton discoveries: "Date unknown, modern era: Northwest California. Three scientific aides and a wildlife biologist from California Department of Fish and Game, District 1, Eureka, California, found the bones of two adults and one juvenile Bigfoot. The FBI was notified and came to take the bones. A judge then issued a gag order on the case, and nothing more was heard. Government coverup. Reported by Ray Crowe."[96] As usual, although intriguing, such vague assertions lead nowhere and cannot be treated as serious proof.

Next up from Lindsay, we read: "Date unknown, modern era: Morgan Lake, Santiam Highway, Oregon. Southwest of Portland, three gigantic skeletons were seen in the lake under four to six feet of water. Mysterious holes had appeared on the ice-covered lake that winter. It was thought that the Bigfoots had used the holes to bury their dead in the ice-covered lake. Reported by Ray Crowe."[97] Morgan Lake is near La Grande, in Union County, and while Crowe's March 2011 blog refers to "three bodies reported in Morgan Lakes [sic]," the IBS database includes no such report.[98]

Crowe's database dishes up more confusion with the next incident from Chelan County, Washington, reported twice under separate case numbers but clearly identical. Item #782, credited to "Chelan prospector Stephen Maher," reads: "DEAD BF FOUND, RED EYES, VOCALIZATIONS." Case #3216, credited to Peter Byrne and an undated, untitled "Wenatchee news article," reads: "[F]ound a dead Bigfoot with red eyes, SCREAMS."[99] As usual with cases from Washington, the database does not deliver the "full reports" promised for either incident. My inquiry to the local *Wenatchee World* newspaper failed to locate the elusive article.

Another case from the database, #2564, comes from Washington's Clark County, credited to Peter Byrne and Larry Dillard. Lacking its "full report," the item simply reads: "HAIRY FOOT FOUND (IT WAS A BEAR PAW)."[100]

The IBS does somewhat better with our last two cases. The first comes from Marion County, Oregon, and reads: "Steve Williams reports an area beyond Abiqua Falls off Road #200 (sic #300?) where there is a cave with stacked bones. The road is a long one; and son Jamie is still looking for the exact area southeast of Silverton; OR. Steve says there are some beautiful waterfalls out that way. He plans to do an extensive survey of the area commencing July 25th."[101] The absence of a follow-up report suggests nothing was found.

Next from the IBS database comes a second-hand account, reported by one Jackson Moore on his sister's behalf. Referring to Clackamas County, Oregon, it reads: "Jackson Moore says his sister knows of a cave on Clear Creek, feeds from Clear Lake, where bones were found at one time. Will try to organize an outing to the area in better weather to see what they could be."[102] Once again, the absence of a follow-up report dashes our hopes.

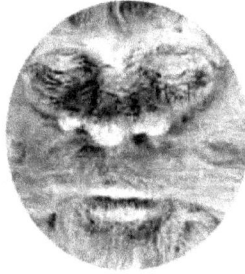

Chapter 9.
Caught Alive?

A living Sasquatch would be even more persuasive than old bones, and we have 10 reports on file of manimals captured by humans from the early 19th century to modern times. They all prove disappointing in the end, but bear discussion here, in our pursuit of the evasive type specimen.

Robert Lindsay leads the list, writing: "1820's: Near Pomona, La Verne and Claremont, California. A 'Devil Indian' or Bigfoot female, was captured by early White settlers, but was soon released. The local Gabrielino Indians reported that the Devil Indians of this area were tall, hairy, smelled bad and roamed around at night. They had large hands and feet and were very fast. Reported by J. P. Harrington."[1] The three towns named are all located in Los Angeles County, but none existed in the 1820s. According to their websites, La Verne and Pomona were settled in the 1830s, while Claremont's first Anglo-European settler arrived in 1871.

Lindsay's source is presumably John Peabody Harrington, an agent of the Smithsonian Institution's Bureau of American Ethnology whom we previously met in Chapter 8, describing skulls found at Santa Barbara in 1923. An admiring website describes him as an "angry god, perfectionist, paranoid worrier, culture hero, obsessed genius, thorn-in-the-side, doggerel poet, ruthless slave driver, inattentive father, valued friend, skinflint, ascetic, academic outcast, great phonetician, indefatiguable [sic] fieldworker, outrageous, laughable and endearing eccentric." The same site says that Harrington collected "close to a million pages of notes on more than 90 different languages, as well as numerous recordings and artifacts."[2] That mass of material may contain some mention of a captured "Devil Indian," but if so, I was unable to retrieve it.

The Bigfoot Encounters website delivers our next item, courtesy of researcher Scott McClean, clipped from the *Boston Times* of April 1, 1839, under the headline "Will Wonders Never

Cease?" Bearing in mind the date of publication—April Fool's Day—and the total lack of any independent corroboration, the piece tells us that on January 21 employees of the New York Western Lumber Company wounded and captured a huge, hairy "wild man" along the Saint Peters River. Also bagged were "two cubs" that proved "active and playful" in captivity. The adult stood eight feet three inches tall, and "his legs are not straight, but like those of a dog and other four-footed animals, and his whole body is covered with a hide very much like that of a cow." Company spokesman Robert Lincoln promised to exhibit the creatures free of charge, "this forenoon, in the rear of No. 9 Elm Street" in Boston.[3] The hoax collapses when we realize that Wisconsin has no Saint Peters River, and that the New York Western Lumber Company apparently never existed.

Bigfoot Encounters provides our next case as well, reading: "21 January 1855—Waldoboro, Maine. J. W. McHenry chopping wood heard loud screams coming from a wooded area near his home. Looking up, he saw a[n] 18 inch creature with black hair. He is said to have captured it and kept it."[4] Said by whom? Janet and Colin Bord repeat the story, calling the witness "J. W. McHenri," citing their source as an article in Olympia, Washington's *Pioneer and Democrat* newspaper, published on May 12, 1855.[5] Thankfully, that article exists online, and is in fact a reprint of J. W. McHenri's letter to the editor of Maine's *Thomaston Journal*. It reads:

> Mr. Editor: On the morning of Jan. 2d, while engaged in chopping wood a short distance from my house in Waldoboro, I was startled by the most terrible scream that ever greeted my ears; it seemed to proceed from the woods near by. I immediately commenced searching round for the cause of this unearthly noise, but after a half hour's fruitless search I resumed my labors, but had scarcely struck a blow with my axe when the sharp shriek burst out upon the air: Looking up quickly, I discovered an object about ten rods from me, standing between two tree[s], which had the appearance of a miniature human being. I advanced toward it, but the little creature fled as I neared it. I gave chase, and after a short run succeeded in catching it.
>
> The little fellow turned a most imploring look upon me, and uttered a sharp shrill shriek, resembling the whistle of an engine. I took him to my house and tried to induce him to eat some meat, but failed in the attempt. I then offered him some water, of which he drank a small quantity. I then gave him some dried beach nuts [sic], which he cracked and ate readily. He is of the male species [sic], about eighteen inches in height, and his limbs are in perfect proportion. With the exception of his face, hands, and feet, he is covered with hair of a jet black hue. Whoever may wish to see this strange specimen of human nature, can gratify their curiosity by calling at my house in the eastern part of Waldoboro, near the Trowbridge tavern. I give these facts to the public, to see if there is any one who can account for this wonderful phenomenon.[6]

If others ever viewed the creature, their impressions have passed unrecorded.

Reports of hairy bipeds roaming Arkansas's Ouachita Mountains date from 1834, and one such—dubbed the "Giant of the Hills"—was allegedly captured in Saline County, sometime after the end of the Civil War. Folklorist Otto Ernest Rayburn describes the mute, nude giant as seven feet tall and covered in long black hair, although he was reputedly "of the white race." Lassoed and forced to don clothes, the captive soon ripped them off and later escaped, eluding pursuers as he—or it—fled toward Texas.[7] With it vanished all corroborating evidence.

A more substantial case, ostensibly, comes to us from the Louisville, Kentucky, *Courier-Journal*, published on October 24, 1878, and subsequently carried in other newspapers as far distant as Galveston, Texas. Dr. O. G. Broyler reportedly captured the "wild man" after years of sporadic sightings around Sparta, Tennessee, and at some point the specimen passed to "third owner" John Henry Whallen, manager of The Metropolitan, a vaudeville theater in Louisville. According to the article, Whallen "promise[d] to successfully baffle all scientists who desire to give a satisfactory explanation of his [the wild man's] unnatural appearance." More specifically, "Close inspection shows that his whole body is covered with a layer of scales, which drop off at regular intervals in the spring and fall, like the skin of a rattlesnake. He has a heavy growth of hair on his head and a dark reddish beard about six inches long."[8]

That sounds very little like Sasquatch—and indeed, it was not. The wild man's original captor, Dr. Broyler, explained that the prisoner was the son of "respectable citizens of North Carolina, named Croslin," who fled the family home at age five and had lived wild ever since, until trapped on September 15, 1878.[9] In February 1879 another physician, Dr. L. P. Vandell of Louisville, diagnosed the subject as suffering from ichthyosis, a skin disease that produces skin resembling fish scales.[10]

"Jacko"
North America's most famous report of a Sasquatch capture premiered on July 4, 1884, in the *Daily Colonist*, published in Victoria, British Columbia. Few readers of *Sasquatch Down* will have missed that story, but I repeat it here for those still in the dark.

> In the immediate vicinity of No. 4 tunnel, situated some twenty miles above this village [Yale, B.C.], are bluffs of rock which have hitherto been unsurmountable, but on Monday morning last were successfully scaled by Mr. Onderdonk's employees on the regular train from Lytton. Assisted by Mr. Costerton, the British Columbia Express Company's messenger, and a number of gentlemen from Lytton and points east of that place who, after considerable trouble and perilous climbing, succeeded in capturing a creature which may truly be called half man and half beast. "Jacko" as the creature has been called by his capturers, is something of the gorilla type standing four feet seven inches in height and weighing 127 pounds. He has long, black, strong hair and resembles a human being with one exception, his entire body, excepting his hands, (or paws) and feet are covered with glossy hair about an inch long. His fore arm is much longer than a man's fore arm, and he possesses extraordinary strength, as he will take hold of a stick and break it by wrenching or twisting it, which no man living could break in the same way.
>
> Since his capture he is very reticent, only occasionally uttering a noise which is half

bark and half growl. He is, however, becoming daily more attached to his keeper, Mr. George Telbury, of this place, who proposes shortly starting for London, England, to exhibit him. His favorite food so far is berries, and he drinks fresh milk with evident relish. By advice of Dr. Hannington raw meats have been withheld from Jacko, as the doctor thinks it would have a tendency to make him savage. The mode of his capture was as follows :

Ned Austin, the engineer, on coming in sight of the bluff at the eastern end of the No. 4 tunnel saw what he supposed to be a man lying asleep in close proximity to the track, and as quick as thought blew the signal to apply the brakes. The brakes were instantly applied, and in a few seconds the train was brought to a standstill. At this moment the supposed man sprang up, and uttering a sharp quick bark began to climb the steep bluff. Conductor R. J. Craig and Express Messenger Costerton, followed by the baggage man and brakemen, jumped from the train and knowing they were some twenty minutes ahead of time immediately gave chase. After five minutes of perilous climbing the then supposed demented Indian was corralled on a projecting shelf of rock where he could neither ascend nor descend. The query now was how to capture him alive, which was quickly decided by Mr. Craig, who crawled on his hands and knees until he was about forty feet above the creature. Taking a small piece of loose rock he let it fall and it had the desired effect of rendering poor Jacko incapable of resistance for a time at least.

The bell rope was then brought up and Jacko was now lowered to terra firma. After firmly binding him and placing him in the baggage car "off brakes" was sounded and the train started for Yale. At the station a large crowd who had heard of the capture by telephone from Spuzzum Flat were assembled, each one anxious to have the first look at the monstrosity, but they were disappointed, as Jacko had been taken off at the machine shops and placed in charge of his present keeper.

The question naturally arises, how came the creature where it was first seen by Mr. Austin? From bruises about its head and body, and apparent soreness since its capture, it is supposed that Jacko ventured too near the edge of the bluff, slipped, fell and lay where found until the sound of the rushing train aroused him. Mr. Thos. White and Mr. Gouin, C.E., as well as Mr. Major, who kept a small store about half a mile west of the tunnel during the past two years, have mentioned having seen a curious creature at different points between Camps 13 and 17, but no attention was paid to their remarks as people came to the conclusion that they had either seen a bear or stray Indian dog. Who can unravel the mystery that now surrounds Jacko? Does he belong to a species hitherto unknown in this part of the continent, or is he really what the train men first thought he was, a crazy Indian?[11]

While superficially persuasive, and still with some believers among Sasquatch researchers, the *Colonist* story was soon deflated by correspondent "Rex," writing from Yale to another newspaper, the *Guardian*, three days later.

The "What Is It" is the subject of conversation in town this evening. How the story originated, and by whom, is hard for one to conjecture. Absurdity is written on the face of it. The fact of the matter is, that no such animal was caught, and how the "Colonist" was duped in such a manner, and by such a story, is stranger, and stranger still, when the "Columbian" reproduced it in that paper. The "train" of circumstances connected with the discovery of "Jacko" and the disposal of same was, and still is, a mystery.[12]

On July 11, New Westminster's *British Columbian* carried the following item:

> The Wild Man—Last Tuesday it was reported that the wild man, said to have been captured at Yale, had been sent to this city and might be seen at the gaol [jail]. A rush of citizens instantly took place, and it is reported that not fewer than 200 impatiently begged admission into the skookum house. The only wild man visible was Mr. Murphy, governor of the gaol, who completely exhausted his patience answering inquiries from the sold visitors.[13]

There, alas, we must leave Jacko, in the realm of fiction from which he came. An odd and wholly unsupported footnote to the story comes from Robert Lindsay, who writes: "There are reports that soon after, a Bigfoot matching Jacko's description was shot and killed in the same general area by a group of men."[14]

"Evidently Truthful"
Two years after Jacko's capture was reported, the *Republic County Pilot* of Cuba, Kansas, carried an even more startling tale from correspondent W. H. Whimple. It is reprinted here, to the best of my knowledge, for the first time since its original publication on September 9, 1886.

THE WILD FAMILY
The Four Critters Captured After a Long Chase
How They Look and Act—A Strange Story, but Evidently Truthful
The Women Caught

> Linn, Sept. 2, 6 a.m.—Reports coming in from Parsons creek say that the wild woman and girl have been caught, and the man surrounded, but it is thought he will not be taken without a desperate struggle. The woman cannot talk, but makes a peculiar noise, something between a grunt and the growl of a dog, which the girl appears to understand, but those about them have no idea of the meaning. This is the report of Mr. F. A. Kingsbury and Mr. Wm. Cummins, who were there and saw them. The elder one was caught some two miles north of Palmer on Peach creek. The younger woman was caught about half a mile farther on the creek, where she was hiding in some hazel brush. They were securely bound, put in the wagon, and taken to the camp ground, and delivered to Sheriff Scott, who will take them over to Washington as soon as the man is captured. I am going over to the camp ground to see them and will try to [come] back in time to send report by this day's mail.

The Man Caught

> Parsons Creek Camp Ground, 8:30 a.m.—I left Linn at 7 a.m., and am here in the

presence of the strangest creatures I ever saw. They are confined in a tent, bound and guarded. The good ladies have supplied them with clothing, and now they are dressed from the waist down. They will not eat cooked food, but devour raw meat much as a dog would. The older woman had a heavy black beard, and long thick hair of the same color, with a bald spot on top of the head about two and a half inches in diameter, which looks like the scalp had been removed. There is also a depression in the skull near the right temple, which looks like it might have been produced with a club. She also has hair on her face, but short and fine, though the hair on her head is long, black and coarse. They have very black eyes, rather flat nose, thin lips, large feet and hands, and are quite muscular, sufficiently so at least to keep the crowd from pressing too close. They are neither white, black, nor red, but of a dirty ash color.

Ben Bartlett has just come in from Linn, bringing the little child, which, when taken into the presence of the women, appeared wild with delight. The older woman took no notice of it whatever, but the younger one as soon as she saw it, commenced making a kind of growling, whining noise, and made desperate efforts to break her bonds, which had the effect to scatter the crowd, especially the female portion.

Rev. Andrew Black and J. J. Patterson, of Linn, who have been out with the searching parties, have just come in and say the man was caught three miles east of Clifton, on Parsons creek, by T. M. Doland's party, after a hard fight, in which Rev. Black received a blow with a club in the hands of the wild man, which broke his arm about three inches above the elbow. The arm was set by Dr. Tyler, of Clifton, who was with the party. Mr. Dolan [sic] and party are bringing the man in and is expected here any time. Quite a number have gone out to meet him. Mr. A. S. Race and Dr. Williamson, of Washington, are here and offer to take charge of the family and provide for them. I suppose they want to put them on exhibition. I do not know what else they could do with them.

The Latest

Dolan is here with his man, who appears stupid and dull. He is perfectly naked, but the people have contributed clothing, which will be put on at once. The man is about the same color as the woman, and has a very heavy beard and long hair. His breast is also covered with hair. He is about five feet ten inches high, and apparently very stout. He has evidently been scalped. The people here think that the man and woman are man and wife, and that some time in the past they were scalped by Indians and became insane from the effects of their wounds and have been wandering in the woods ever since. There is strong talk among the doctors here of performing an operation on the woman's head and see if they can restore her mind sufficiently to find out something of their history. The operation will be performed to-morrow, if at all.[15]

And there the story ends, without an epilogue—at least, until Robert Lindsay described it in 2011, writing: "September 5, 1886: Washington County, Kansas. Four Bigfoots, a male, a female, a young female and a juvenile, were captured. They were covered in black hair and could not communicate. The female only made grunting sounds."[16]

The original report, however, is not so clear-cut. Aside from ashen skin tones and the adult woman's "heavy black beard," the captives generally displayed human characteristics, being regarded as a pair of traumatized and mentally defective adults raising two feral children. We have no measurements for their "large" feet and hands, and the absence of a spoken language, at least with the young ones, is common to feral children. Kansas logged its last uprising of Plains Indians in October 1868, although a minuscule Scott County "battle" claimed one life in September 1878, 216 miles southwest of Linn, eight years before the wild family was captured,.[17]

A Sasquatch family, or hapless victims of the Midwest's brutal race war? Either way, their fate is lost to history.

Wild Stragglers

Robert Lindsay floats our next story, briefly summarized as follows: "February 15, 1908: McHenry, North Dakota. An apparent Bigfoot was captured near town. He was covered with hair and had eye teeth like fangs. He refused to eat and could not communicate. He only drank water, half a bucket at a time. Reported by the Stevens Point Journal, Stevens, Wisconsin, February 15, 1908."[18] Reports carried in out-of-state newspapers, sans local coverage, sometimes signaled journalistic hoaxes in the 19th century. To cover the respective bases, I sent inquiries to the *Stevens Point Journal,* and to libraries in both McHenry County, North Dakota, and to the town of McHenry, in Foster County. No source could produce the article cited by Lindsay.

Three decades passed before the next alleged Sasquatch capture, reported on Bigfoot Encounters.

> Bristol Bay, Alaska: 1940—Near the ghost town of Kaluka, Alaska.
>
> Emily Supanich's mother was berry picking with others when they came upon a large hairy creature that resembled a man covered with long black hair. They ran back to the village and told the people. The men went out, captured it and caged it. She said her mom fed it raw fish. After some time the hair began to fall out and it turned out to be a female with breasts. Not long after the hair started falling out, the creature died. This is recorded in a letter written to Roger Patterson by Emily Supanich, San Bruno, CA. John Green's cardfiles, BC Archives.[19]

I found records of an Emily T. Supanich residing in California, age 69 in 2014, but was regrettably unable to contact her before press time for *Sasquatch Down.* If the same person, she would have been born five years after the reported incident, and would have no personal memories beyond the story she was told in childhood.

Our last purported Sasquatch capture dates from December 2012, announced by Darren W. Lee, executive director of the Mid-America Bigfoot Research Center, "Where researchers think outside the box."[20] Lee's message, posted at 9:27 a.m. on December 28, under the heading "Did it happen?" posed the question: "Did the Quantra group obtain a live specimen with their operational plan to capture a Bigfoot?"[21]

The "Quantra group" was scarcely known to other researchers before December 27, when self-described "former member Ed Smith" allegedly received "an automated text message from the group's messaging system." That message read:

> December 27, 2012 10:09 AM
> From: CINC—6
> As of 0906 27 DEC 2012, "Daisy" is in the box.[22]

Mr. Smith went on to say, "I believe this message is factually correct because when we referred [to] BF while under observation we used the term 'target.' The designation was to change to 'daisy' once captured. This was never published or posted. 'CINC' is capture control and '6' is the chief or lead. No one else would have known this. Plus the routing and IP information matches. I believe that Quantra has a live specimen. If this is correct then I congratulate them on a job well done."[23]

Another message from Smith to Lee, also posted on December 28, claimed that Daisy had been sedated and moved to an "examination area" 12 miles from the undisclosed capture site, where "it could take up to 72 hours to conduct the examination. A press release is expected within the 48 hours. It appears that an unprecedented event is in motion, having been on the inside of this operation and now observing from the outside is a defiant [sic] change."[24] Smith had no description of the creature, "nor the actions leading up to the capture of the specimen," but he went on to write (uncorrected):

> Here is what I'm speculating: the examination team is continuing to assemble, examination should take 72 hours.
>
> If they go by the plan then a decision about release or storage in a repository will be made with in 48 hours after the examination is completed.
>
> If release is chosen then a press release should be forth coming after the examination is competed, if the specimen is sent to the repository then a press and information release would happen with in 30 to 90 days thats by the operations plan.
>
> Reasons for moving the specimen to a repository include health issues, prolonged examination and on the darker side private investment group interests.
>
> For clarity, the Original six was never part of the MABRC and the Original six ceased operations completely. Quantra was formed with elements of the O-6, new members, ownership and management. I'm posting this due to inaccurate reporting by some Internet news shows...
>
> There is an old Army saying "once the first round is fired all plans go out the window" I commend the Quantra Group for maintaining operational focus, I guess

my advise on hiring ex-military operators was heeded.

In my opinion, there should be a press release within 48 hours if we get it I would be surprised, it seems they are playing this to the ink as it were.
I will be in touch.[25]

Before day's end on December 28, Rob Gaudet—a founder of Squatch Unlimited and Louisiana state director for the MABRC—weighed in online to say:

We are familiar with the organization that is claiming to have captured the bigfoot, QUantra, formerly ORIG-6, and with Ed Smith, the person who has posted about it in the MABRC forums. Thus, Squatch Unlimited is uniquely positioned to separate the facts from the fiction. We will be posting updates on this as we receive them....
This and all news about this incident IS originating from Ed Smith, a former member of the QUantra group...Ed says that he has since received further information from a member insider indicating that the creature was being moved and will be reviewed by individuals in the Bigfoot and scientific community with the proper credentials. They will decide on what to do with it once it has been examined. We can also say that it may have been captured inside of a large box type trap when it stepped on a trigger device located inside of the trap.

Who is Ed Smith?

Ed Smith was a former member of the ORIG-6 group, which is now known as QUantra. Ed has been a long time poster to the MABRC forums. To be clear, Ed Smith is not a MABRC organizational member. Ed is however a regular poster on the MABRC forums, which are open to the general public, thus MABRC is only a conduit for this information from Ed. Ed's posts were always relevant to the research of the ORIG-6 group's findings. It is believed that Ed lives in Oklahoma, I am unable to confirm that with the forums down.

How is MABRC Involved?

As stated before, Ed Smith has been a long time contributor of the MABRC forums, he is not a MABRC Organizational member. Having been involved in the MABRC forums for several years, Ed has built up trust with the MABRC founders and has posted various interesting findings over the last half decade that he's been on the forums. MABRC is only posting what it knows directly from Ed and is urging patience.[26]

Speaking for the MABRC, Darren Lee granted an interview to blogger Melissa Adair, posted to the Bigfoot Chicks website before year's end. Lee told Adair that the "Original 6" group formed "about 12 years ago," with "a 10 year mission to collect as much evidence as possible to prove the existence of Bigfoot. After the 10 years, Mr. Smith became burnt out and decided to leave the group. The group then renamed itself 'Quantra'." The group was self-funded by members including "doctors, lawyers, engineers, tech heads, etc. They are men of substantial means and able to fund a serious operation out of pocket." Daisy, Lee said, had been caught in a "special trap, a box, with a pressure plate in the center. The plate was set to go off at 350 pounds. It closed up around the creature when it stepped on the plate." A "privately contracted scientist" was on tap to study the creature, and Lee ruled out Melba Ketchum, as "Quantra is not a fan of Dr. Ketchum." All forthcoming information would be published, Lee said, on the MABRC's public forum.[27]

Nothing more was heard until January 2, 2013, when researcher Shawn Evidence posted a video clip on YouTube, summarizing his knowledge to date. In that two-minute presentation, Evidence noted that Steve Kulls—former Tom Biscardi aide, then critic (see Chapter 5)—had joined Rob Gaudet to "verify" Ed Smith's credibility. Smith, meanwhile, named an otherwise unidentified "Mr. Alexander" as a principal member of Quantra, also referring to Dr. Jeff Meldrum, a professor of anatomy and anthropology at Idaho State University, as "an associate of this collection." The latter claim seemed iffy, as Smith also said Quantra hoped for a future meeting with Dr. Meldrum "if available." Evidence closed with the announcement of a three-hour documentary film "in production.[28]

Steve Kulls chimed in next, on January 7, to announce creation of a "bridging group" to examine Daisy in the flesh. Suggested by Darren Lee, said group would include a team from the MABRC (Lee, Rob Gaudet, Doug Todd, Randy Harrington, and Jim Whitehead), plus an "independent" group of non-MABRC members including Kulls, Dr. Meldrum, Kathy Strain, Melissa Hovey, and Abe Del Rio. Professing caution, Kulls wrote, "My personal 'proof of life' is to lay eyes on their specimen. No more, no less, videos, photos, interviews with Team Quantra will not suffice.... I will be releasing no further information on this independently, as I am now part of an assembled group. However I will not allow this to be a long, drawn out process either. This is not for an air of secrecy, but to prevent any wild speculations or conclusions."[29]

Another week elapsed before Darren Lee pulled the plug with his final pronouncement online. On January 14, 2103, he posted the following message, condensed below but otherwise unedited:

> The last few weeks has seen perhaps the biggest hoax ever perpetrated has occurred. Ed Smith has stated publicly on the Bigfoot Forums that he has hoaxed the MABRC for the last 4½ years.... Let's get something straight right off the bat, the buck stops here with me. Regardless of the speculation and assumptions, no one from the MABRC knew that Ed was hoaxing.... I tracked down the names of other members of the so-called "Orig-6" and found that they were real people, whether they were actual friends of Ed or not, we couldn't verify.... The database that I created on Ed contained every thing he had ever told Randy [Harrington] and I, along with his posts, we did this to see if we could find discrepencies, and the stories did not change. We could not crack his stories to prove whether he was making it up or tellin the truth.... No one will ever know what all has transpired in this entire affair, I don't have the time to sit down and explain everything to people. The MABRC and it's researchers conducted this like a witness investigation, and the only way to see if it was true, was to go the distance.... Now that the Quantra Affair has come to an end, Ed seems appalled that everyone is now digging deep into his

personal life, and to attempt to throw the investigation of himself into obscurity, he begins claiming he hoaxed the MABRC, hoping to put the heat on the MABRC instead of Ed himself.[30]

In short, it was Georgia 2008 all over again, without even a cheap rubber costume to show the public. Invoking Dr. Meldrum's name was also part of the charade; he writes, "I had no association with the quantra group and the whole thing was a scam that hardly garnered any of my attention."[31] Motives remain obscure, but the stench of fraud lingers like Sasquatch musk in a dank forest glen.

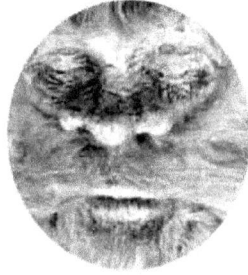

Chapter 10.
Manimals Abroad

N orth America is not alone in its long history of Sasquatch-Bigfoot sightings. From aboriginal prehistory to modern times, every inhabited continent on Earth has produced reports of hairy bipeds at large, and with those sightings come reports of creatures killed or captured. This chapter, while by no means exhaustive, shall touch on some highlights of the global search for evidence of manimals.

Zana

The Caucasus version of Sasquatch is known as the Almas, sometimes rendered as Almasty. Around 1850, a female Almas was allegedly captured by residents of T'khina, a village 50 miles from Sukhumi in the Ochamchiri District of Abkhazia, Georgia. Villagers named the creature Zana ("black" in Georgian), gradually taming her, teaching her to perform manual labor—and, on the side, to sexually service one Edgi Genaba, described in some accounts as a local nobleman. Genaba repeatedly impregnated Zana, and while several of their children died in infancy, at least four survived, given away by useless father Edgi to neighboring families. Survivors included son Dzhanda (born in 1878), daughter Kodzhanar (1880), daughter Gamasa (1882), and son Khwit (1884).[1]

Zana died in 1890 and was buried in an unmarked grave, but locals in T'khina kept better track of Khwit, deceased in 1954. A Russian yeti researcher, Dr. Igor Burtsev, unearthed Khwit's skull in 1971 and subsequently gave it to a group of anthropologists in Moscow, who professed themselves "amazed" by the relic's seeming mixture of "primitive" and "progressive" (i.e., modern) characteristics.[2] One panelist, M. A. Kolodieva, said the skull "approaches closest the Neolithic Vovnigi II skulls of the fossil series."[3] Two decades later, in the early 1990s, Dr. Grover Krantz pronounced the skull entirely normal, lacking any Neandertal features.[4]

There matters rested until 2013, when Dr. Bryan Sykes, a professor of human genetics at the

University of Oxford, conducted DNA tests on Khwit's skull and samples from six of Zana's surviving descendants, announcing a startling result. Khwit's ancestry—and, by extension, Zana's—was found to be 100 percent Sub-Saharan African, perhaps the offspring of slaves imported under the rule of the Ottoman Empire.[5] That made nonsense of the early tales describing Zana's capture in the wild, and laid to rest forever any question that she may have been a stray Almas.

Ameranthropoides loysi

In the midst of World War I, European nations took a sudden interest in the natural resources of South America, particularly the rich oil reserves found in Venezuela. In 1917, Swiss geologist Dr. François de Loys joined the hunt for black gold, employed by Standard Oil—but allegedly found something quite different. In addition to fending off savage natives, whom he blamed for killing 17 members of his party, de Loys also claimed that he had met—and killed—the "missing link" between humanity and apes. According to his story, while exploring territory west of La Fría, capital of García de Hevia Municipality—

> ...The jungle swished open, and a huge, dark, hairy body appeared out of the undergrowth, standing up clumsily, shaking with rage, grunting and roaring and panting as he came out onto us at the edge of the clearing. The sight was terrifying...

> The beast jumped about in a frenzy, shrieking loudly and beating frantically his hairy chest with his own fists; then he wrenched off at one snap a limb of a tree and, wielding it as a man would a bludgeon, murderously made for me. I had to shoot.

> My Winchester got the best of the situation. Riddled with bullets, the great body soon fell on the ground almost at my feet, and quivered for a while. He gathered his arms over his head as if to hide his face and, without a further groan, expired.[6]

De Loys examined his kill, reporting that it had no tail, a trait common to great apes but not to monkeys. It also had 32 teeth, he said, four less than a New World monkey's normal complement of 36. De Loys named his discovery *Ameranthropoides loysi* ("Loys's American ape"), and allegedly preserved its head and hide. Alas, both relics were lost before he returned to civilization, during sundry misadventures and battles with hostile tribesmen. The only evidence of his discovery was a single black-and-white photo.[7]

That snapshot was problematic at best. It shows the creature seated on a wooden packing crate of unknown size, held upright by a stick of unknown length propped underneath its chin. Today, most viewers of the photo think its subject is a common spider monkey (genus *Ateles*) with its long prehensile tail concealed by the full-frontal camera angle. De Loys himself would not broadcast the story until 1929, initially published in London's *Illustrated News*, then reprinted by the *Washington Post* under the headline "Found At Last—the First American."[8] In print, De Loys said:

> Until my discovery of the American anthropoid, we could only imagine that man migrated to these shores. But now, in the light of this discovery, it is obvious that

the failure of the otherwise well established principle of evolution when it was applied to America was due only to imperfect knowledge. The gap observed in America between monkey and man has been eliminated; the discovery of the Ameranthropoid has filled it.[9]

De Loys and his creature owed much of their celebrity to French anthropologist George-Alexis Montandon, a racist fascinated with eugenics, author of the 1940 publication *Comment reconnaître le Juif* (*How to Recognize the Jew*), whose collaboration with Nazis in occupied Paris prompted his assassination by members of the French Resistance in August 1944.[10] Montandon eagerly promoted De Loys's "missing link" theory in 1929, using *Ameranthropoides loysi* as the centerpiece for a new theory of human evolution in the Western Hemisphere. The net result, given Montandon's white-supremacist proclivity, was a hypothetical connection between Native Americans and De Loys's ape, but most scientists rejected both that conclusion and the Latin name proffered for the alleged type specimen.[11]

After World War II, with both de Loys and Montandon deceased, the story began to unravel. Author Earnest Hooten, in his book *Man's Poor Relations* (1946), recalled a conversation with James Durlacher, an American geologist who investigated de Loys's claim in 1927. Hooten wrote: "At this time, he made inquiries of men who had been in the de Loys expedition, and discovered that the specimen de Loys shot was, indeed, a spider monkey, known in that region as the marimonda." Fifteen years later, in a 1962 letter to the Caracas newspaper *El Universal,* Dr. Enrique Tejera wrote, "De Loys was a bromista ["joker"], and many times we laughed with his jokes. One day he was offered a monkey. It had its tail sick, and it cut it off itself. De Loys called it 'el hombre mono' [the ape-man]." According to Tejera, when the pet died from natural causes, De Loys snapped his famous photo for posterity, not in the jungle, but outside the city of Mene Grande.[12]

French cryptozoologist Michel Raynal found earlier confirmation for Tejera's view of De Loys in a 1960 book by Raymond Faisson, *Des Indiens et des mouches* (*Indians and Flies*). Speaking of Dr. Tejara, Faisson wrote: "He told me that, while physician for the Mene Grande Oil Company, near Maracaibo, he knew that Loys had staged a photograph of a dead spider monkey near the camp. The proof, he said, was the presence of a foot-tall banana plant visible to the background of the original. Bananas were introduced in America and cannot grow in the wild in the unexplored forests of the Upper Tarra."[13]

Blogger Brian Dunning poked more holes in De Loys's story, in 2012, noting that the alleged native attack that left 17 members of his party mortally wounded occurred within 10 miles of La Fria. "How likely is it," Dunning asked, "that a geology party would allow seventeen men to die without simply making the short return trip to La Fría or some other town? How likely is it that de Loys would have continued prospecting if any men had actually been violently killed on the job? The job simply wasn't worth men's lives, and at no time was de Loys in so remote a position that he could not easily have returned to civilization. In short, his story, as printed, is almost certainly a gross exaggeration in the style of the popular adventure fiction of the day."[14]

That said, de Loys was not the only outsider spinning tales of apes or apemen killed in the vicinity. American engineer, diplomat, and amateur ethnologist Richard Ogelsby Marsh penned the following account in 1934, describing alleged events in Panama's Darién Province, bordering northern Colombia.

> I saw In 1920 while I was in Panama an old and experienced American prospector, Shea by name, came to me with a strange story. He had just returned from a trip to southeastern Darién. With another American he had ascended the Sambu River, which enters the sea on the southern shore of San Miguel Bay. The country here was, and still is, wholly unknown. Even the mountain range back from the coast did not appear on the maps. Shea and his companion worked their way with great difficulty to the headwaters of the Sambu and there they became separated. The other American has not been heard from since.
>
> When Shea lost his companion, he lost his canoe and most of his equipment. So instead of attempting to return down the Sambu River, he decided to forge his way to the Pacific Ocean across the Andean Range to the west. He reached the divide in a state of exhaustion and by a stroke of luck stumbled on an old Indian dugout abandoned on the bank of a small river running into Pinas Bay. It was nearly dark, so he camped for the night at a considerable altitude not far from the divide.
>
> All that night he heard the footsteps of a large animal in the jungle above his camp. And when dawn came, he heard a curious chattering sound. He looked up and saw standing on top of the bank an animal that appeared to his unscientific mind to be a cross between a Negro and a gigantic ape. It was six feet tall, walked erect, weighed possibly three hundred pounds and was covered with long black hair. It was glaring down at him and chattering its teeth in rage.
>
> Shea whipped out his revolver and shot it through the head. It tumbled down the bank and lay still beside his canoe. When Shea recovered from his fright he measured the animal crudely. It was heavily built like a gorilla, but the big toes on the feet were parallel with the other toes, as in a human being, not opposed like thumbs, as in all other monkeys and great apes.
>
> Unfortunately Shea was too exhausted to bring any part of the animal back to civilization. He barely managed to get down to Pinas Bay on the Pacific and attract the attention of a coaster, which took him to Panama more dead him many times after that in the hospital where he eventually died of chronic malaria. Almost his last words were a solemn oath that the story of the "man-beast" was true.
>
> Of course, my first reaction to this story was extreme skepticism. But I found to my surprise that many trustworthy men who had penetrated into the little-known parts of tropical America did not share my disbelief. The "man-beast" is reported to have been seen in many locations.
>
> A Spanish gold-hunting expedition in the seventeenth century reported that it had shot fourteen of the Man Beasts not far from this same Pinas Bay.[15]

Fact or fiction? Once again, without a specimen, the mystery endures.

Almas Again

Russia's revolution of 1917 did not bring immediate peace and stability to the former tsarist empire. Civil war raged between "Red" and "White" troops from November 1917 to October 1922, and anti-Bolshevik guerrilla action continued on through the 1920s. In 1925, Soviet Major General Mikhail Stephanovich Topilski led an assault on insurgents holed up in a cave, in the Pamir Mountains of Tajikistan's Gorno-Badakhshan Autonomous Province. One of the surviving rebels told Topilski that hairy apemen had attacked his squad before Russian soldiers arrived on the scene, and presented a carcass to prove it. Topilski wrote:

> At first glance I thought the body was that of an ape. It was covered with hair all over. But I knew there were no apes in the Pamirs. Also, the body itself looked very much like that of a man. We tried pulling the hair, to see if it was just a hide used for disguise, but found that it was the creature's own natural hair. We turned the body over several times on its back and front, and measured it. Our doctor made a long and thorough inspection of the body, and it was clear that it was not a human being.[16]

Understandably, Topilski's mission did not permit preservation of the unique specimen.

As a wider war enveloped Asia in the latter 1930s, and the storm clouds gathered over Europe, more reports of Almas killed or captured reached the West. Dordji Meiren, a member of the Mongolian Academy of Sciences, claimed to have seen an Almas pelt in 1937, in a Buddhist monastery at Baruun Haral, in the vast Gobi Desert.[17] As summarized by author Myra Shackley:

> It was being used by the lamas as a ritual carpet for some of their ceremonies. The hairs on the skin were reddish and curly, and the skin had been prepared after death by cutting it down the spine, so the face and chest were preserved. The features were hairless, the face had eyebrows, and the head still had long disordered hair. Fingers and toes were in a good state of preservation and the nails were similar to human nails.[18]

Again, no samples were obtained for scientific testing.

Three years later, in 1940, Mongolian troops assigned to guard the Manchurian border against Japanese invaders allegedly ambushed a party of suspected infiltrators, killing several. On closer inspection, the bullet-riddled corpses were not Japanese, but hairy bipeds of some indeterminate species. No trophies of the hunt were kept for subsequent examination.[19]

Around the same time, in 1940, Chinese biologist Wang Zelin filed the following report while working with the Yellow River Irrigation Committee:

> Around September or October, we were traveling from Baoji to Tianshui via Jingluo City; our car was between Jingluo City and Niangniang Plain when we suddenly

heard gunshots ahead of us. When the car reached the crowd that surrounded the gunman, all of us got down to satisfy our curiosity. We could see that the "wildman" was already shot dead and laid on the roadside. The body was still supple and the stature very tall, approximately 2 meters [6 feet 6 inches]. The whole body was covered with a coat of thick grayish-red hair which was very dense and approximately one *cun* [1¼ inches] long. Since it was lying face-down, the more inquisitive of the passengers turned the body over to have a better look. It turned out to be a mother with a large pair of breasts, the nipples being very red as if it had recently given birth. The hair on the face was shorter. The face was narrow with deep-set eyes, while the cheek bones and lips jutted out. The scalp hair was roughly one *chi* [1 foot] long and untidy. The appearance was very similar to the plaster model of a female Peking Man [*Homo erectus pekinensis*]. However, its hair seemed to be longer and thicker than that of the ape-man model. It was ugly, because of the protruding lips.[20]

In December 1941, Russian soldiers reportedly captured an Almas-type creature 18 miles north of Buynaksk, a town located in the foothills of the Greater Caucasus on the Shura-Ozen River, in what was then the Dagestan Autonomous Soviet Socialist Republic. Since German troops had invaded Russia six months earlier, nervous Soviet soldiers suspected the creature— locally known as a Kaptar—might be a disguised Nazi spy. They delivered it to Lieutenant Colonel Vazhgen Sergeyevich Karapetian of the Red Army Medical Corps, whose notes describe the prisoner as nearly six feet tall, covered from head to feet with shaggy brown fur resembling a bear's. Karapetian pronounced the captive a "wild man," and later heard that, although innocent of any crime, it had been executed by a firing squad.[21]

Myra Shackley cites one more incident, involving discovery of a drowned Yeti near the village of Tharbaleh, close to southern Tibet's Rongbuk Glacier, describing the creature as "about the same size as a small man with a pointed head but covered with red-brown hair." Viewing the creature as an omen of bad luck, villagers quickly disposed of it.[22] Shackley offers no date or source for the story, but authors Michael Cremo and Richard Thompson date the incident from 1958, while citing Shackley as their source.[23]

Inscrutable
Throughout Nepal and Tibet, various Buddhist monasteries harbor supposed Yeti remains, generally scalps or desiccated hands, with occasional undocumented claims of whole pelts seen by transient witnesses. In most accounts, the relics are considered sacred, though some monasteries have been known to show them off to tourists, for a price. If even one such item was legitimate, forensic scientists might answer, once and for all, the question of the "Abominable Snowman's" existence.

That thought preyed on the mind of Texas oilman and adventurer Thomas Baker Slick Jr. in the 1950s, as he bankrolled expeditions to find the Yeti, Sasquatch, the Loch Ness Monster, and a supposed giant salamander lurking in California's Trinity Alps. For Yeti evidence, Slick focused on a hand preserved by lamas at Pangboche, in Nepal's Solukhumbu District. Slick photographed the hand in 1957, but since snapshots prove nothing, he determined to obtain a piece of the relic itself. Team member Peter Byrne drew the assignment, removing several bones from the hand and replacing them with human bones, smuggling his prize across the border into India, there transferring the package to film star James Stewart, vacationing at the time in Calcutta. Stewart, in turn, used his celebrity status to slip the bones past Customs officers to the West. An alternate version claims that Pangboche monks sold one of the hand's bony fingers to Byrne, being complicit in replacing it with human bones. Ironically, when Sir Edmund Hillary and Marlin Perkins visited Pangboche in 1960, the hand's human finger convinced them that the relic was a hoax.[24]

Meanwhile, London University primatologist William Charles Osman Hill examined the bones Byrnes had obtained, pronouncing them a fair match for *Homo neanderthalensis*. Three decades later, in 1991, author Loren Coleman discovered that American anthropologist George Allen Agogino had acquired and retained some fragments from the Pangboche sample. Analyzed for the NBC television program *Unsolved Mysteries* in February 1992, the bones were vaguely identified as "near human." December 2011 brought an apparently final verdict from the Royal Zoological Society of Scotland in Edinburgh, where DNA analysts declared the Pangboche relics human in origin.[25]

Sometime after the 1992 NBC broadcast, thieves stole what remained of the Yeti "hand" from Pangboche, along with a supposed Yeti scalp—or replica skullcap, stitched from goat's skin. That raid left the monks short of items to draw tourist dollars. In April 2011, New Zealand mountaineer-adventurer Mike Allsop announced that he had commissioned replacement relics from Weta Workshop, whose special effects team created most of the creature makeup for film director Peter Jackson's *Lord of the Rings* trilogy. "I will take these replicas back to the monks so they can replace the ones that were stolen," Allsop told reporters. "I want to help the monastery have an income again. I want to help them out."[26]

Finally, in the same month that the Royal Zoological Society of Scotland rendered its DNA verdict on the Pangboche bones, reports reached the West of a Yeti-type creature captured in the Republic of Ingushetia, a federal subject of Russia located in the North Caucasus region with its capital at Magas. Border patrol agents allegedly caught the hairy female biped after a shepherd chased it away from his flock. Early descriptions sketched the captive as six feet six inches tall, omnivorous, with arms shorter than its legs. Alas, the story proved to be another hoax, staged for publicity on New Year's Eve.[27]

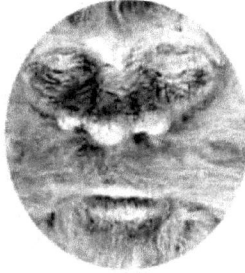

Chapter 11.
Black Helicopters

T he annoying disappearances of rumored Sasquatch corpses, bones, and gravesites may be explained in one of four ways. First, the solution preferred by "skofftics" maintains that since no such creatures exist, all persons who claim to have killed one or found one's remains are mere hoaxers. A second and more charitable possibility suggests that witnesses reporting Sasquatch bones or bodies in the wild have glimpsed some other large mammal, either too damaged or too decomposed to be identified by anyone but an expert. Third, we might grant that the size of Sasquatch, the terrain in which it normally appears, and other factors could prevent a man—or group of men—from harvesting and packing out remains.

The fourth solution, and the subject of this chapter, is conspiracy.

This theory, popular with certain Sasquatch researchers, suggests a wall of silence consciously erected by the "Powers That Be" in North America, including local, state and federal law enforcement agencies, the U.S. military, the Smithsonian Institution, plus various other museums and universities. Robert Lindsay, summing up the conspiracy thesis in 2011, writes:

> For the past 43 years, there appears to be a government coverup about Bigfoots. Over a 43 year period, there were 16 cases of government coverup of Bigfoot evidence. That is one coverup case every 2.7 years, about one incident every 3 years. Before 1968, there were no government coverup cases. The evidence suggests therefore that there may be a government coverup dating from 1968. In particular, government officials have been taking Bigfoot bodies away, never to be seen again. The government appears to be involved at various levels, including the National Guard, the Army Corps of Engineers, state police, the FBI and the US military.[1]

In fact, we have one anonymous claim of a Sasquatch-concealment plot occurring six years prior to 1968, dating from October 10, 1962. On that day, the Bigfoot Hotspot Radio Facebook page says, "a[n] elderly female Bigfoot, suffering from Lyme Disease, estimated age 40, about 8'2 and weighing 770 pounds," died when a large Douglas fir fell on top of her during a storm near O'Brien, Oregon. Members of a road-clearing crew allegedly found the decomposed carcass on November 9 and summoned U.S. Forest Service (USFS) officials, who arrived in force at noon. An unnamed USFS district manager confiscated photos of the creature, though the anonymous author assures us that "some photos in private hands still exist. For many years, one of the photos was on display in a local restaurant." The carcass was removed, aboard a truck, and "Three days later, the local ranger, Guy Adams from Cave Junction, began spreading a lie that the Bigfoot was a pet ape that had escaped from a local residence."[2]

So far, the tale has all the classic elements of a conspiracy theory: vanishing evidence, rumored items that escaped the sweep but can't be found today, a cover story, and even the name of a real-life forest ranger tossed in for confirmation. (Ranger Guy Adams was assigned to Montana, also covering parts of Idaho and the Dakotas, in August 2014.[3] He ignored my email inquiries.) Having described the creature's hush-hush handling, the blog strangely proceeds to chart the corpse's subsequent movements in great detail, from Oregon to California, then to Golden, Colorado, and finally to a San Diego facility run by the U.S. Department of the Interior. According to the unknown author, "A file on the Bigfoot, USFS/33058-45333-294734-19B, along with bones and photos, was sent to the Smithsonian Museum, then the file case was closed to the public, and the photos and file were marked classified. The DOI ordered the Smithsonian to send the bones to the DOI on May 24 1963."[4]

We might ask why the DOI, already in possession of the Sasquatch's remains, needs to demand return of bones from "the Smithsonian." That conundrum pales, however, when the blogger proceeds to quote from the alleged file, which he says "is stamped 'classified,' and marked exhibit 4377," reading, in part:

> Species: unknown biped
> Date recorded: 3/14/63
> Area: O'Brien-Dew Ridge
> The bone structure of the specimen is unknown to DGDS
> Analyses: Tissue samples indicate non human.
>
> Regarding unknown biped. The subject discussed in the original file is complete with the finding of Dr. D. S. Gould. This is a medical conundrum as to the true species of said subject. Subject appears to be some species not known to date. Some indications are most related to human. Yet many indicate of a gorilla type. It is noted the length of the subject is clearly not gorilla nor human because subject measurements indicate 98 inches in height. Estimation weight at time of death 770 lbs. This clearly concludes this subject is not consistent with known species of human or gorilla.
>
> Conclusion: Sample is not consistent with any known species of animal/primate known. Seal per request noted.[5]

The blogger goes on too say: "The USFS could not figure out what the animal was. The feeling that it was some sort of an ape, granted it was bigger than apes typically are. The report did not indicate that the animal was a Bigfoot, since the USFS didn't know what that was at the time. In 1964, the DOI classified the Bigfoot as very similar to an Eastern China Mountain Gorilla. [N.B.: No such species exists.] They thought it was an escaped pet. Much of the remains were destroyed through testing and over time, so there may be few if any left. Reported on the Bigfoot Ballyhoo blog. However, some say that this whole story was made up by Linda Newton-Perry, who is a pathological liar."[6]

The final gratuitous, possibly libelous, comment refers to a well-known Sasquatch researcher who, with husband Christopher Perry, owns the Bigfoot Ballyhoo website. Steve Kulls, discussed in Chapters 5 and 9, refers to Newton-Perry (no relation to this author) as a hoaxer on his Squatchdetective website, listing various examples of supposed misleading information posted to Bigfoot Ballyhoo[7], while the Bigfoot Lunch Club site apparently supports her activism in the field.[8] Whichever view is accurate—if either—the 1962 remains and "classified" report remain elusive.

William Jevning, previously met in Chapter 3, offers our next second-hand report of an alleged conspiracy in Oregon. The tale begins in autumn 1966, when an unnamed email correspondent went hunting with his parents at China Hat Butte in the Deschutes National Forest. A large nocturnal prowler frightened the father and left him vowing he would never return to the butte, but a year later he (the father) began "a secret letter-writing exchange" with his twin sister, plotting a family vendetta against Sasquatch for the scare it had delivered. They planned the campaign for deer season in October 1968, but the correspondent's dad suffered an aortic aneurysm that September and died 10 days later. Before his passing, on a morning walk to school, Jevning's correspondent saw a U.S. Forest Service vehicle parked near his house, unoccupied. Returning home that afternoon, he "checked the letter collection stash file, [and] the letters were gone. All I've ever thought is that the Law Enforcement arm of the USFS broke into our home and took the letters. All without any warrant presented to our family! How would they have found the location? One way, only one way, and that was to bully a man laying dying in a hospital, my father Harry! After the death of my dad, the private contract postal station in our suburb of Keizer treated my mother and myself very rudely! Was there a USFS Law Enforcement intercept of our mail going on?"[9]

The Lindsay Files
Robert Lindsay's "official" roll call of Sasquatch cover-ups begins in 1968, with the first case including a spin-off. He writes:

> 1968: North of Carson, Wyoming. Three men were hired by a rancher to kill a Bigfoot that was killing his cows and sheep by tearing off their legs. Afterward, the body was picked up by a government helicopter and taken to a research facility in Almogordo [sic], New Mexico to be autopsied and studied. Government coverup. Reported by Ray Crowe.[10]

> After 1968: Alabama. The same man involved in the Carson, Wyoming case above

shot another Bigfoot later on. This time the government found out about it and was angry that he killed the Bigfoot. Reported by Ray Crowe. Government coverup.[11]

The IBS database includes no cases resembling those cited by Lindsay for either Wyoming or Alabama, but Ray Crowe did mention the events in his March 2011 blog, writing: "Dennis [?] said there was a Bigfoot problem and he and two friends were hired to hunt down the Bigfoot that was killing cows and goats by pulling off their legs. This was north of Casper, Wyoming, and they shot it dead. The body was picked up by helicopter and supposedly transported to an Alamogordo, New Mexico, research facility to be autopsied and destroyed. How did he know that? And why not Area 51? Later his companions killed another Bigfoot in Alabama, only this killing the government objected to. I wonder why?"[12]

Without identifying "Dennis," we shall never know.

Leaping forward eight years and across the continent to the East Coast, Lindsay offers the following item:

> June 1976: Baltimore, Maryland. As unlikely as it sounds, a Bigfoot was reported here in May 1976. Police were called, and K-9's initially refused to track it. Finally, the dogs tracked it to an interstate tunnel. A police officer then saw it run under the interstate. The next month, US army personnel were called out to deal with the Bigfoot. Reports indicate that soldiers captured or killed the Bigfoot. No further information. Government coverup. Reported by Rick Berry, Bigfoot on the East Coast.[13]

Berry's book does allude to soldiers capturing a Sasquatch in Maryland, sometime during June 1976, but he offers no source or supporting details.[14]

In 1997 researcher Bobbie Short received the following letter from one Craig Bennett, otherwise unidentified, whose common name made him untraceable during my research for *Sasquatch Down*. Bennett wrote:

> Back in December 1977, some twenty years ago, I was a freshman at Burlington County College located in Pemberton, New Jersey. Next door is Maguire Air Force Base and Fort Dix Army base. So it was a common sight see to the military in the college courses. While sitting in the snack bar and trying out my first cup of coffee, a soldier walked in and went over to a table and then a second soldier arrived going to the first one and excitedly asked him if he had heard about the news. He related that a platoon of soldiers on an exercise from (I'm not sure of the fort name—Fort Lewis, Washington?) went into the woods to drill and were injured in an encounter with Bigfoot somewhere in the Pacific Northwest. The first soldier then looked around the room nervously and then [at] the other soldier and told him to not talk about it here and "we'll do it later." I did not see them again nor do I know their names.[15]

Short surmised that Bennett's letter might relate to a case briefly discussed in Chapter 4.

Readers will recall the tale of soldier Edwin Godoy, stationed at Fort Lewis in Washington State, who allegedly shot a Sasquatch with red "self-luminous" eyes while on night patrol, sometime in 1978. The creature escaped, but left footprints and bloodstains that seemed "strangely oily." Godoy reported the incident to other soldiers, and—

> From that moment on they kept at a distance and wouldn't talk to him. They communicated by radio to the base and reported the incident... At about 7:30 A.M. some unknown personnel arrived to the site: several men dressed in white lab coats, wearing thick gray "rubber" (leaded?) gloves and boots took samples from the tracks impression on the ground, the alleged "blood" which was handled with extreme care... Later, they all were ordered by radio to return at once to Ft. Lewis. Godoy was to report himself to the base hospital immediately at his arrival.

> To his surprise, an Air Force medical officer, a colonel, was waiting for him there. Fort Lewis is a U.S. Army military base with no ties with the Air Force, so why the presence of this full-bird Air Force colonel there? He couldn't say. The usual thing would have been for the regular medical staff in the base hospital to attend him. This man was not from the hospital's medical staff. The officer debriefed him thoroughly on the incident and made a complete medical and physical exam to him. While examining him he kept asking at what distance he was from the creature when he shot at it, on the creature's description, if he felt a tingling sensation or had a sore throat, headaches, if a rash had developed on his skin — and other things. The Air Force medical officer apparently knew what to ask. It was obvious to Godoy that he was looking for specific symptoms — and answers — but symptoms and answers to what?

> Several samples of his blood, skin scrapings, urine, saliva and other types of samples were taken from Godoy. The soldier knew something odd was going on, he kept asking the officer where he had come from but he wouldn't answer. After being examined, he was ordered to go to his barracks, then he took a shower and rested. Later, he was ordered to go to the base commander's office. The base commander, (a lieutenant general—name not remembered by Ed Godoy), was there together with his company commander, Captain Underwood, and a colonel whose last name was, to his best recall, Kropsie. They debriefed him again on what had happened out in the woods and then the base commander ordered Godoy not to talk ever to anyone on what had happened. He was warned that if he ever talked about it he'd be court martial[ed] and would have to face the consequences.[16]

Soon afterward, Godoy was approached by "L. Robles, a Puerto Rican soldier who was commissioned in the [base] hospital's lab." Robles questioned him about the shooting, then said, "I, together with two other guys, had to analyze the blood samples taken from the ground, and we know you are the soldier involved because it was stated as such in the report... And you know? It's crazy, but... what the hell was it you shot out there? When we examined the blood samples we found out three weird things in it... That blood contained human blood cells, animal blood cells...and chlorophyll. Man, that's incredible! What the hell was it?"[17]

Godoy had no idea, but "thinking back, he feels that the base commander, Colonel Kropsie, and Captain Underwood, all seemed to know what they were dealing with, and for that reason they had ordered him to keep his mouth shut on the incident. He remembered that on one occasion he had to enter a huge security vault in the base in which many bottles are stored. All these bottles were filled with a liquid substance that had a greenish glow, similar to what Robles had described. The bottles in the vault were kept there under very heavy security, because, according to him, the liquid in the bottles seemed to be plutonium stored at the base."[18]

Strange stuff indeed—and, of course, served up without a vestige of supporting evidence.

Craig Woolheater, co-founder of the former Texas Bigfoot Research Conservancy, posted our next case to the Cryptomundo website in October 2005, quoting a story told online several years earlier by a Texas physician, Dr. Pat Sullivan. According to Sullivan, in 1979 he was called to the scene of an auto accident 30 miles northeast of Dallas, police reporting that a motorist had struck a "large bear," propelling it through her windshield, trapping her inside the car. Sullivan wrote:

> Both were still alive when I got there. The place was covered with military and police vehicles and the "bear" was emitting great screams. I was not allowed to approach the vehicle until [the] bear was removed. A police officer shot the animal several times before it stopped screaming. The scream was definitely not bearlike—in fact it sounded almost human. They kept me and all other civilians about a hundred feet away while the military personnel covered the "bear" with a large black tarp and loaded it into a capped truck. One could se the outline of the body under the tarp – it had very long legs, both front and back and was shaped like a very large man. I asked to see the animal before they took it away. They treated my request with a cold "No!" and told me to mind my medical business. The woman was unhurt but in shock. She kept saying that what she hit was not a bear but a big hairy man and others gathered around were saying that it was a bigfoot. I never saw anything about the accident in the local papers.[19]

Woolheater was unable to contact Sullivan for further details, pegging him as "one of the ones that got away," but a physician of that name died in Bedford, Texas, at age 83, on January 19, 2008.[19]

Thar She Blows!

On May 18, 1980, volcanic Mount St. Helens erupted in Washington State, claiming 57 human lives and causing an estimated $1.1 billion in property damage ($2.88 billion today), depositing ash in 11 states and five Canadian provinces. The titanic explosion occurred at ground zero for Pacific Northwest Sasquatch sightings, and it should be no surprise that rumors have emerged describing manimals killed or injured by the blast. The IBS database includes seven cases of Sasquatch corpses allegedly seen in the wake of the eruption, and while none includes the promised "full reports," their brief summaries provide a fair overview of the claims.

Report #994: 2 BF CORPSES FROM MT ST HELENS CORP OF ENGINEERS DREDGED.

Report #1015: DEAD BF BODIES FROM MOUNT SAINT HELENS SEEN.
Report #2063: BF BODIES FLOWN BY COPTER, BURNED.
Report #2616: DEAD BF SEEN IN HELICOPTER NETS FROM MT ST HELENS ERUPTION.
Report #3467: CORPSES OF BF FROM MT ST HELENS.
Report #3468: TWO BF BODIES FOUND AFTER MT ST HELENS, CHOPPER TAKES THEM.
Report #3493: DEAD BF FROM MT ST HELENS SEEN.[20]

As previously noted, the database is well known for duplicate and even triplicate entries of a single case, and without further details we have no way of knowing how many actual sightings those seven listings represent. Ray Crowe helps in that respect, with his blog of March 2011, detailing five specific incidents. Taken in order of appearance they include an unnamed witness in Spokane who "saw a double rotor helicopter with a big star pass between 100 and 150 feet overhead, with hairy arms, legs and disguising debris covered with volcanic ash hanging out of the cargo net"; witness "Jamie's" claim that "a crane dredging operation on the Cowlitz River by the Manatowaka Company found two bodies buried in the sand after the blast. The bodies were found two weeks after the blast and choppers flew them off. Jamie says a crane operator said while doing cleanup, he saw several unmarked choppers with nets hanging below them, arms and legs dangling out. He was cautioned to say nothing"; "Terry's" assertion that "the Army Corps of Engineers took out two Bigfoot corpses two months after the St. Helens blast"; a report from "Fred's dad," a security officer at a National Guard camp near Bremerton, who "saw military helicopters bring large numbers of animals and Bigfoot bodies in cargo nets into the compound. They were doused with napalm and burned, totally destroying the bones and carcasses"; and another witness to that burning who "was told to keep his mouth shut after seeing the Bigfoot creatures."[21]

Two months after Crowe posted his blog, Robert Lindsay offered a revised or garbled version, reading:

July 1980: Mt. St. Helens, Washington. Up to 20 dead Bigfoots were dredged out of a river after the volcano eruption. They were taken by helicopter to a place where all the dead animals from the eruption were being buried in a landfill. The Bigfoots were guarded by National Guard troops. Asked what was to be done with them, the troops said that authorities were probably going to study them. A truck came to take the Bigfoot bodies away. Five different witnesses, one named Fred Bradshaw, reported that Bigfoots were discovered by searchers and hauled away by arms of the government, including the Army Corps of Engineers and the National Guard. Witnesses were warned to keep quiet. Government coverup.[22]

Tales also exist of Sasquatches surviving the eruption, then winding up in human custody. Robert Lindsay summarizes one such story: "1980: Yakima Indian Reservation, Washington. Several Yakima Indians noted that a Bigfoot had been badly burned in the Mt. St. Helens eruption but had managed to survive. It was hanging out on the eastern end of their reservation afterward. They reported that a government helicopter came, bundled up the wounded Bigfoot and took it away. Government coverup. Reported on the Phantoms and Monsters website."[23]

In fact, the website referenced was quoting a letter sent to researcher Shawn Evidence, purportedly written by a former Washington National Guardsman who demanded anonymity. The letter's author claimed he was assigned to cleanup duties after the eruption, during which "Myself and four other guardsmen were told to follow a group of soldiers and not to speak to each other and to remain very quiet overall. We were told to get into a jeep and wait. We sat in the jeep for maybe a half hour. Eventually another jeep arrived carrying a civilian and another member of the military. The civilian was brought into the tent and he emerged a few minutes later followed by a large hairy creature. It looked like a large man covered in fur and the best way to describe it was like 'Beast' from X-Men only brown. The creature looked to have some burns and had a bandage on its arm. At first we were afraid but when it walked by we could see its eyes and it just looked very sad and somber. He climbed into the back of a pickup with the civilian and the two were speaking in a weird language I had never heard. It would cough at times."[24]

Guardsmen followed the truck to five different areas, where the injured creature called out toward various caves. At the second and fifth stops, creatures "just like the one with the civilian" emerged from hiding, both suffering from burns, and were helped into the truck, then returned to the base camp. As the first creature returned to its guarded tent, "it looked at us and made a waving gesture with its hand. We took it as a thank you for what we had done." The Guardsmen were debriefed by a high-ranking officer who allegedly told them, "These creatures live in these areas; they mean no harm and want to be left alone. Do you really want to do anything that may cause them trouble? They are like us in a lot of ways. If you need or want to talk about this just wait about 30 years, by that time there will likely be no reason to keep them a secret."[25]

Shawn Evidence also posted the story of an unnamed airman allegedly stationed at California's George Air Force Base in 1980. When Mount St. Helens erupted the airman's unit was engaged in training exercises with troops from British Columbia's Canadian Forces Base Comox on Vancouver Island. During a training flight, their plane malfunctioned and was forced to land at Travis Air Force Base in northern California for repairs. While there, they were issued ill-fitting uniforms with "GUARD" stenciled on them, together with M16 rifles, then posted to sentry duty around the air base. The letter's author was assigned to guard a tent, where he saw purported victims of the St. Helens eruption escorted inside by military personnel. He observed that "the escorted 'people' were REALLY wide, but I always assumed (until now) that it was the coats there were wearing." After perusing the National Guardsman's story online, the airman revised that opinion, suspecting that he had seen captive Sasquatches, although nothing that he witnessed proves that case. Before leaving California, the visitors were cautioned to keep silent about their brief interlude at Travis AFB.[26]

In May 2010, interviewed online by blogger Scott Brown, Loren Coleman opined that the stories swirling around Mount St. Helens constitute "the worst kind of rural folklore, because no one has ever been found who was the primary source of that story.... Really, the military could care less. They're not looking for Bigfoot, they're looking for spies, terrorists, even illegal immigrants. It's not that the government and military services are stupid, it's that they have only so much brain power and memory, both technically and legally. That is closer to the truth. It's not an overt cover-up."[27]

Others, of course, beg to differ.

Through the Grapevine

Robert Lindsay presents our first post-eruption tale of conspiracy, from the same neighborhood: "Unknown date, probably after 1980: Cowlitz or Yale, Washington. Near Mt. St. Helens, a Bigfoot was killed by a vehicle on a highway. A US Forest Service helicopter came and took the body away. Nothing further was heard. Government coverup. Reported in the Bigfoot Track Record [sic]."[28] In fact, this incident—like many others—pops up twice in the IBS database, under separate case numbers.

Report #1138: BF KILLED BY CAR, FOREST SERVICE CHOPPER TAKES BODY AWAY.

Report #3905: BF HIT BY CAR, CHOPPER FLEW IT AWAY.[29]

Neither account includes the promised "full report."

In his next item, Robert Lindsay writes: "1982: Klamath, Oregon. On the Winema National Forest southeast of Crater Lake National Park, a Bigfoot was killed by a car. US Forest Service and local police sealed off the area, and the Bigfoot was hauled off in a truck. Government coverup. Reported in the Bigfoot Track Record [sic]."[30]

A supporting document from the IBS database reads: "J. W. said his brother-in-law had been a witness where a Bigfoot was run over by a lady driving near Bend, OR, close to Frog Camp, in 1982. She said that the Forest Service, county and city police were involved and sealed off the area off Hy. 97 where the incident occurred. Supposedly a truck carted off the covered body, but the feet were sticking out. THERE IS A FROG CAMP IN LANE COUNTY, FAR FROM HWY 97...BUT A FROG LAKE IN KLAMATH COUNTY. THINK REPORT IS BOGUS."[31]

The IBS report refers to Frog Camp, an historical place in Lane County. There is also a Frog Lake Campground, located in the Mount Hood National Forest off Highway 26, in Clackamas County. Statewide, Oregon has six Frog Lakes in five counties (Curry County having two).[32] Wherever the event allegedly occurred, no official record of it remains.

Our next case takes us to the Midwest, where Robert Lindsay writes: "1984-1985: Martin County, Indiana. Dogs cornered a female Bigfoot as she was giving birth. She ran away, abandoning her fetus. The dead fetus was taken to the game warden, but all evidence disappeared after that. Government coverup. Reported by the Gulf Coast Bigfoot Research Organization website."[33] The GCBRO's account, logged by an unnamed correspondent claiming multiple Sasquatch sightings, includes a statement that "We don't talk about it much here but my grandfather and uncles were hunters and kept hunting dogs. One [Sasquatch] gave birth and the dogs cornered it. The fetus was turned over to the local game warden and all evidence of it has disappeared. My granfather's [sic] dogs were slaughtered."[33]

Another GCBRO report describes members of the U.S. Army Special Forces pursuing a Sasquatch around Maricopa County, Arizona, in May 1987. The anonymous witness, one of four civilian campers who allegedly observed the hectic chase, claims he shot the creature when it eluded the soldiers but failed to kill it. (He also confuses his weapon, oddly calling it a shotgun and a rifle in different paragraphs.) Afterward, he wrote, "Nobody believed us when we reported our encounter to the police and game warden and here is the funniest part. Two days later the game warden tried to arrest me for shooting an endangered species. When I told him, 'You arrest me but first you have to tell me what it was I alledgedly [*sic*] shot,' he then backed down from arresting me but told me to stay where I could be found for further questioning. "[34]

Robert Lindsay returns with our next tale of conspiracy: "1991: Chelan, Washington. On the east slope of the Cascades, US Forest Service officials photographed a dead Bigfoot that was found by a prospector. Government coverup. Reported in the Bigfoot Track Record [*sic*]."[35] The IBS database includes 14 reports from Chelan County, but none resembles the story related by Lindsay.[36]

Fire on the Mountain
On August 6, 1998, simultaneous wildfires swept through Nevada's Lander County, threatening the county seat at Battle Mountain. One day later, a supposed federal government firefighter penned a letter to the BFRO that read:

> I observed an animal wounded by fire moving on all fours not like a bear. More like ape. Fire fighters captured animal, contacted local vet and medical doctor. U.S. Department of Fish and Wildlife, Department of Interior, and Bureau of Land Management on the scene.
>
> Animal tranquilized and moved to unknown location. Those at scene told not to talk about what they saw.
>
> Animal approximately 7.5 feet long/tall, human like arms and legs, face not like man or ape but mixed between. Genitalia: male, uncircumcised and human-like. Hair covering most of body except chest, chest has hair but sparse, hands with sparse hair, palms bare, with five digits with human opposition of thumb and 5th digit.
>
> Speech—attempted to communicate with care-givers once it realized they were attempting to care for it. Multiple burns on hands, feet, legs and body; some 2nd and 3rd degree burns, using "rule of nines" approximately 45 per cent of body with burns.
>
> Doctor and Vet working together providing care and moved it to unknown location locally. This notice given in violation of orders given by BLM, DOI and DF&W. Witnesses numbered in the area of 30-25. Word is out in the government agencies, and among the firefighters, since an M.D. was called out. Many thought a firefighter was injured.
>
> Please note that I am a government employee of one of the listed agencies fighting brush fire in wilderness area of Nevada (large scale fire approximately 70,000 acres burned) and under orders not to disclose information.

I believe a cover up is in the making, people need to know, the animal needs to be kept alive and studied and released in protected area.[37]

Posting that letter online in July 2011, Shawn Evidence quipped that "not even Robert Lindsay can top this story," but he was mistaken. Lindsay had already added a twist to the tale, two months earlier, in a blog that read: "The Bigfoot was taken to a university in the Bay Area, where it was treated for a few days, then returned to an area about 150 miles from where it was captured. Government coverup."[38] Lindsay lifted that tidbit from Thom Powell's book *The Locals*, recounting Powell's telephone conversation with the anonymous witness, claiming that after the creature's injuries were treated, "it was moved to an undisclosed location, against orders of Bureau of Land Management, Department of the Interior and Department of Fish & Wildlife." That smacks of conflicting conspiracies, but we shall learn no more, apparently, since the witness—known to Powell only as "Marty"—has broken off contact.[39]

The Game's Afoot
Another possible military pursuit crops up in this short item from Dennis Bauer: "1999-09-00; AL, unknown; a reported bigfoot sighting and an armed helicopter shooting into woods."[40] Robert Lindsay elaborates.

September-October 1999: Northwestern Alabama. A woman sighted a Bigfoot and reported it to law enforcement. The next day at 4 PM they saw helicopters flying over the area. The copters had .50 caliber machine guns and were firing into the woods. This went on until midnight. Apparently they hit the Bigfoot because residents heard horrible screams from the wounded Bigfoot. It is not known if the Bigfoot was killed or not. When people asked law enforcement about the helicopters, citizens were told that the police had been eradicating wild boars in the area. However, there had not been any wild boards [sic] in the area for 20 years. Reported by the [now defunct] Southeastern Bigfoot Research Organization. Government coverup.[41]

Lindsay's next two reports allege cover-ups in two road-kill cases reported in Chapter 6, one from witness "Clarissa," the other investigated by Ray Crowe and his wife. Neither of the original stories involves any official meddling, but Lindsay still regards them as incidents in an ongoing conspiracy.

July 2000: 13 miles east of Hood River, Oregon. Along the Colombia [sic] River, a dead Bigfoot was spotted on the highway on Interstate 84 off the highway in the westbound lane, apparently killed by a vehicle. The Bigfoot was grey and was stretched out in a fetal position with an arm outstretched. There were several witnesses. No further data on the case. There were no newspaper reports. Government coverup.

2001: Hood River area, Oregon. Another Bigfoot was reportedly killed on Interstate 84. When investigators went to investigate, they found only tracks and no body. Government coverup.[42]

On January 20, 2001, two anonymous friends riding all-terrain vehicles found apparent Sasquatch tracks in a creek bed near Point Pleasant, West Virginia—notorious for 1960s "Mothman" sightings and other strange events over the years. The property's owner, identified only as "Jack," notified the state's Division of Natural Resources, whereupon "they sent one truck down to look at the print. This truck left and came back with about five more trucks. They told Jack not to come back there while they were there. After they were done one of them stopped and told Jack that it was only a black bear print. I am no zoologist but I have seen black bear prints, and this was not one. There were no claw marks, but there were toes. And if that was a bear, and it left tracks that big, I wouldn't want to meet it!"[43]

In March 2005 the Oregonbigfoot.com website received an anonymous second-hand report of a Sasquatch found dead near Duvall, Washington, 26 miles east of Seattle. According to the unnamed correspondent, relating a tale from a friend identified only as "XXXX":

> Basically, what happened was XXXX found a dead Bigfoot on his property. He has 3 or 4 German Shepherds and they led him to the creature. XXXX got a real up-close look at the thing, obviously. He said it was face-up and had no obvious wounds or injuries. XXXX reckoned that perhaps the dogs scared it to death, but he didn't really know why it was dead. He described it to me as being around 7-8 feet long, maybe 350-400lbs. in weight. The thing that seemed to impress him the most was the things hand-size which he said 3 times bigger than his, the fingers being particularly long, the fingernails were black. It was very hairy he said. Now here the weirdest part. XXXX said he phoned-in to the Police Department (I am assuming the Duvall Police). He said the police never showed-up, but with-in an hour or so a black twin-rotor helicopter landed. Men in black uniforms came-out and ordered XXXX to stay in his house. XXXX said the men spread-out a tarp, then rolled the creature onto it, picked it up, and put it in the helicopter and flew-off with it. End of story. All I can say about the yarn is I know of no reason for XXXX to lie or make-it up.[44]

Terror in Tennessee

Panic gripped Campbell County, in the Norris Highlands of northern Tennessee, during October 2003. Best known for coal mining, the county was aghast at reports of a hairy primate at large, suspected in the disappearance of more than 100 domestic animals. "I didn't really get a good look at his face because he didn't hang around all that long," witness Donna Keathley told reporters. "But he's big and he's got a really bad stinkful odor to him." From the county seat at LaFollette, animal control officer George Moses said, "I believe that the people believe that they see it. I don't have any reason to doubt them, that they're mistaken."[45]

Describing her encounter with the beast, Keathley said, "This kitten he threw at me, no one knows where it come from. If I had a gun, I would've shot it." Another witness, unnamed, glimpsed the creature on October 14 and pronounced it an orangutan, although those Asian apes dine primarily on fruit, insects, and birds eggs. Only nine cases describe orangs consuming smaller mammals, all of the victims slow lorises.[46]

While Campbell County's sheriff questioned registered owners of exotic primates, self-described cryptozoologist Chris Dotson contacted Knoxville's Channel 6 News on October 22, branding the hairy suspect a "skunk ape" of the kind often reported from Georgia and Florida. "The animal people reported seeing is not a pet," Dotson declared. "We need to capture this thing the best we can. People need to stay away from it, allow experts to come in who have some knowledge of large primates like this and trap this animal."[47] Five days later, Dotson and John Placyk, a biologist from the University of Tennessee, collected hair and feces samples from the scene of one reported sighting, announcing plans for DNA testing.[48]

Then, in mid-November, Loren Coleman's Cryptomundo blog broke the news that sheriff's officers had reportedly killed one shaggy prowler, while another escaped. As summarized by Coleman:

> Word out of Campbell County is that on Wednesday, 11/12/03, at about 3 P.M. EST, sirens and gunshots were heard. Screams of the recently sighted apelike animals were also heard at the same time. Sounds were coming from the right side of one of the houses that has been involved in the series of sightings, down toward the lake, near LaFollette.... A member of the Sheriff's department said that they were working with a group to eliminate these animals. This information came forward today, on 11/13/03. Also, on earlier [sic], on Wednesday apparently, two of these cryptid animals were seen in a field. One was large, the other smaller. This was about 9:15 A.M. The shootings happened after that.[49]

Coleman also posted a message from controversial Tennessee researcher Mary Green, stating that a local resident had seen the body of some unidentified black creature lying on her property as she left for work, while aircraft circled overhead. She called her husband to inspect the corpse, but when he arrived 10 minutes later it was gone.[50] In the end, police denied killing or capturing any mysterious creatures in Campbell County. Predation on pets and small livestock abated, and no DNA test results were ever published on the samples collected by Chris Dotson and John Placyk—all of which spelled conspiracy to Robert Lindsay, who wrote:

> November 12, 2003: Lafollette, Tennessee. A creature had been killing peoples' animals. A woman's goat and cat at the very least had been killed. She called the sheriff's, and they came out with a team of deputies and told everyone to get their pets indoors as they were going to eliminate these animals. They tracked the Bigfoot and shot it dead over the next hill. There were sirens wailing, and the Bigfoot screamed as it was shot. The woman left the scene. People saw a black body lying in a field the next morning. Ten minutes later, it had vanished. Planes flew around the area night and day for two weeks. Locals reported that there had been a hostile Bigfoot in the area, and they were trying to appease it by leaving food out for it so it would not kill their animals. Government coverup.[51]

Forty years after Watergate, a quarter-century beyond Iran-Contra, eleven years and counting since a concerted campaign of lies sent us to war in Iraq, some Americans have reached the point where government denials of any event count as proof an event has occurred. No evidence, it seems, is needed anymore.

Smoke and Mirrors

Ray Crowe reported our next case in March of 2011, writing: "A Bigfoot, Chiye-tanka, was shot near Slim Buttes on the Pine Ridge Reservation in the Black Hills, South Dakota. The body was sent to the School of Mines Lab to be studied and later given ceremonial rites by three Lakota elders."[52] Two months later, Robert Lindsay pegged the event as occurring in August 2006 and branded it another government cover-up.[53]

Crowe is presumably referring to Rapid City's South Dakota School of Mines & Technology, founded in 1855, which boasts 16 academic departments including a Department of Chemistry and Applied Biological Engineering. This does not appear to be an institution geared toward study of unidentified animal species, being geared primarily toward biomedicine, biopharmaceuticals, biological fermentation, and fabrication of artificial organs.[54] Better choices in the Mount Rushmore State might be Vermillion's University of South Dakota, with departments of biology and anthropology, or Black Hills State University in Spearfish, with its advanced biology curriculum. In any case, it hardly matters, since no school statewide admits to receiving Sasquatch remains.

Our last case with a date attached emerged from Pennsylvania in May 2013. On May 14, Altoona resident John Winesickle called police to report finding "proof of Bigfoot"— specifically footprints—on his Spruce Street property. Before phoning the cops, Winesickle tried the Pennsylvania Game Commission, but its agents failed to call him back. An officer responded to the call, and Winesickle showed him the tracks, which were also photographed. The patrolman suggested they were bear tracks, Winesickle disagreed, and the matter was officially closed... except on the Web, where it quickly acquired a life of its own.[55]

At 2:41 p.m. on May 14—some four hours after Winesickle's call to police—the Pennsylvania Bigfoot Society received its first anonymous email report of the incident, reading: "Today in Somerset county PA a turkey hunter shot and killed an animal he claimed is a Sasquatch. The state police were called and responded to the scene, according to chatter on the local police frequencies the officers confirmed there was an unidentified animal shot and killed. Details are a bit fuzzy at this time."[56] Fuzzy indeed, since Winesickle lived in Bedford County and had reported no shooting.

Matters quickly went from bad to worse. Forty minutes after the first email, another reached the PBS, bearing the author's name and phone number. It claimed a Sasquatch had been killed either on Russell Road or Rustic Road, going on from there to say: "Hello, YES this is legit, a strange report came over the scanner here in Somerset, Pa. just a few miles from Flight 93 memorial.... It was about 20 minutes later [after Winesickle's call to police] that he said he heard a police officer radio back and state, 'there was a body'." After that, the correspondent said, a neighbor heard thunderous sounds overhead. "The sound was so loud it shook his house, so he went outside to take a look. It was then that he observed a formation of four Army Apache helicopters approaching from the north and moving in the direction of Somerset."[57]

Bemused by the strange twist his footprint report had taken, John Winesickle still believes a

Sasquatch family inhabits the woods near his Bedford County home. Although he's never seen the creatures, he describes the male as "a giant" and reports a female "cooing" in the forest. "I think she's pregnant," Winesickle told Somerset's *Daily American.* As for the state Game Commission, "They don't want any part of this." [58]

Dates Unknown

Conspiracies are all the more mysterious when stripped of dates and other details that might aid investigators. Ray Crowe leads the list with a report reading: "Ron [?] heard that the Forest Service Department was doing some clearing of BLM (Bureau of Land Management) land and found a live juvenile Bigfoot in a shelter while it was asleep. They roped it and loaded it into a truck that went to The Dalles, Oregon. He did not know what happened to the creature after that. Possibly dissection? We know where, Nellis of course."[59] The last comment refers to Nellis Air Force Base in southern Nevada, closely associated with the nearby Nevada Test and Training Range that includes "Area 51," rumored repository for crashed UFOs and ground zero for all manner of secret experimentation.

And speaking of Area 51, we find the following anonymous report on the IBS database.

> Caller wants to remain anonymous. Continuing conversation, my caller said he had a friend that had a son that was stationed at Area 51, Nellis AFB, NV, as a security guard with a high clearance, fairly recently. The son said (privately and in confidence to his dad) he was once called into a room where people were dissecting Bigfoot creatures. There were several complete corpses in vats of liquid similar to formaldehyde (which might explain why they didn't need any more bodies from Mt. St. Helens...they had plenty!). There were body parts; hearts, livers, and such, scattered all over on tables. Also he told dad he once got a peek into another room he was passing when the door opened after a retinal scan of another guard...more vats, only containing the "little white aliens with the big black eyes." Scuttlebutt was that the Roswell, New Mexico UFO crash was true, and that the dead aliens and parts of the spacecraft were brought to that installation.[60]

Next, Crowe cites a case discussed in Chapter 8: "Bill [?] said three scientific aides and a biologist of Fish and Game, Region 1, Eureka, found bones of two adults and a juvenile Bigfoot. The FBI was notified and took the bones. A judge declared a gag order on disclosure of the find. That's strange."[61] Of course, the gag order ensures that researchers will find no evidence of the event. As is, we must consider it a rumor, nothing more.

Another item from Crowe's list of undated incidents reads: "Talking to an Ohio Sheriff's deputy about a Bigfoot sighting, the deputy said that he was not surprised at all, as he knew of several sightings in the county himself. He even knew of an incident where a Bigfoot was struck and killed by a car on a State Route. The deputy found a large black hairy man, not a bear, lying alongside the road. He called it into the Sheriff's Department, and soon the Ohio State Patrol came and blocked off the road. Then a dark colored moving van arrived and three men in military outfits put the body in the back of the van and nothing more was heard of the incident."[62]

Next up, another from Ray Crowe: "The Billings Gazette, Montana, reported a Bigfoot shot, or possibly run over by a car—rumors varied. It was on the Crow Reservation and the corpse was supposedly zipped up in a body bag and taken away by federal agents. This was near Pryor, some 40 miles south of Billings."[63] Following Crowe's lead, I contacted *Gazette* editor Darrell Ehrlic and obtained the following article from files Ehrlic describes as "spotty at best."[64]

A Pryor Creek rancher said he didn't shoot bigfoot this summer.

Contrary to widely circulating rumors, Steve Kukowski said no bipedal primate ever terrorized his livestock, was killed by his bullet or was taken away by federal agents.

"The rumors have been going around all summer," Kukowski said. "I have no idea what it is. I have no idea how it got started."

The sasquatch story has persisted for three months, but there are many different versions. Rudy Drobnick, a retired wildlife biologist with the Utah Department of Natural Resources, has been tracking the tales. Drobnick tracks bigfoot sightings across the nation.

"The reports range from an enlarged black widow spider to a giant mouse to a blown up tire on the road," said Drobnick, in a telephone interview from his office in Salt Lake City. "I have no details on any aspect of it other than a bigfoot had been killed, shot or maybe hit on the road somewhere near Billings."

Given the three-month lifespan of the rumor and the emergence of common themes, Drobnick believes further investigation is needed. "At this point in time, based on my information, the thing is to be pursued as if it was a real creature," he said.

In most of the stories, the sighting occurred on the Crow Reservation, the beast was black and hairy and its corpse was zipped into a body bag and whisked away by federal agents minutes before local authorities could investigate.

Dexter Fallsdown, director of public safety for the Crow Tribe, described what he has heard: "A bigfoot got killed, some people came and picked him up and took him away."

Fallsdown said there has never been substantiating evidence.

"We tried to do a follow up on him, but all of the sudden it was hush hush," he said. "I think it's just the rumor mill."

Federal and state agents claim no knowledge of a missing link living near Pryor, about 40 miles south of Billings.

Dan Vierthaler, FBI senior resident agent in Billings, was asked Monday about possible FBI participation in a bigfoot capture near Billings.

"That's the first I heard of it," he said. "I can assure you that we are not doing anything covertly to hide a body. I do not have a bigfoot in my evidence locker."

Vierthaler said he was pleased to be asked about something other than anthrax.

"This could be the next episode of the 'X Files'," he joked, referring to the popular Fox Television show about two FBI agents who investigate the paranormal, including space aliens, vampires and wildly inbred families.

Denials are to be expected from federal agents, Drobnick said. "If there's a dead one, like they said, you wouldn't be able to get anything out of them if you pried it with a crowbar."

Montana Fish Wildlife and Parks, which tracks alien species—including rainbow trout and spotted knapweed—claims no involvement. "I have no comment," Information Officer Bill Pryor said, jokingly. "The FBI cleaned everything up."

If a sasquatch community exists in Yellowstone or Bighorn counties, FWP would be interested in learning more, Pryor said.

"Depending on the size of the population, there's a chance we could trap, tag and redistribute some in other parts of the state," he said. "You know our routine on anything other than humans—try to build them up to huntable populations and start issuing tags."

Drobnick said he and a group of friends have recorded about 100 possible bigfoot sightings over the last 20 years. During hunting season, when Drobnick measures trophy game animals for the Boone and Crocket Club, he asks hunters if they spotted any unusual wildlife in the backcountry.
Most people say no, but there have been credible reports of sightings, Drobnick said....

David Turns Plenty Sr., a minister and Pryor's elected representative to the Crow Tribe's executive committee, said the bigfoot stories continue to circulate in his community.

"There's different stories, but the one I heard was this one rancher had shot it when he saw it amongst his cattle," Turns Plenty said. "They say they took the thing to Bozeman, then I heard again that they took it out of state. A lot of people say it's for it real [sic]. I don't know."

Although no evidence supports the existence of a Pryor ape man, Drobnick said he will keep an open mind and continue searching for proof of the unknown species.
"We'll certainly keep attempting to get to the bottom of this mystery," he said. "If this thing is a hoax, then there's thousands of people being fooled by it. If it's real,

we're just being eluded."[65]

Ray Crowe's next vague report reads: "While on the weird topic, Fate Magazine, December 1990, had a long article of a possible government cover-up when something big was seen in a body-bag at Aberdeen Proving Ground, Maryland."[66] I contacted *Fate,* which no longer stocks that particular back issue, but received no answer prior to press time for *Sasquatch Down.*

In April 2014, Michael Krein wrote to the Bigfoot newsgroup online, describing an experience he had while seeking Sasquatch in northern California, years earlier. I present it here with his permission:

> Back in 2003 when I broke my leg and 6 rangers had to hoof me out of the Siskiyou Wilderness, there was a youngster among them and after I answered his question about what I was doing there he told me that just a few days earlier he had seen a 9 footer cross the main highway through Happy Camp at 2 AM. If looks could kill he would have died on the spot from the glares the other and older rangers gave him. By their reactions, not because they thought he was making it up, but because he was not suppose to admit such things to someone like me. They had obviously heard and maybe discussed his story among themselves before he confessed to me. I don't know if this code of silence is written down somewhere or simply verbally enforced, but it is real and I've seen it exercised first hand. The kid didn't say another word after that.[67]

Why Hide Sasquatch?

Skeptics frequently ask why government, at any level, would strive to conceal the existence of Sasquatch—or, by extension, any other cryptid. Possible reasons suggested by Sasquatch conspiracy theorists include:

1. *Economics.* Verification that the creatures exist might well place them on the federal Endangered Species List, thereby impacting corporations that bank billions of dollars per year despoiling what remains of the environment through logging, petroleum drilling, construction, and so on. History tells us that the exploiters—claiming they support "wise use" of undeveloped wilderness areas—bitterly contested listing of other endangered species such as the spotted owl, snail darter, and others. Indeed, they snarl, mutter, and bribe legislators wherever "tree huggers" attempt to save any scrap of wilderness from corporate despoliation.

2. *Religion.* A Gallup poll from May 2014 indicates that 42 percent of all U.S. residents believe the creation fable outlined in Genesis, while denying the reality of evolution. Among those who claim to attend church weekly, that percentage jumped to 69 percent.[68] Sixty percent of registered Republican voters believe God created all life and matter in its present form within seven days some 10,000 years ago, while only 38 percent of Democratic Party members share that opinion.[69] At a time when right-wing Republicans control the U.S. House of Representatives, dominating its Committee on Science, Space and Technology, the influence of ultra-conservative "pastors" holds sway over the party leadership. The existence of Sasquatch

and other hypothetical related species would make mincemeat of creationism—and potentially destroy some billion-dollar "megachurches" with a huge investment in promoting superstition over science.

3. *Politics as usual.* Finally, there is the natural bent of all governments to "classify" and conceal any "sensitive" information that can feasibly be withheld from the general public. Daniel Benoit, founding leader of the East Coast Bigfoot Researchers Organization, writes that "the Government is simply a bunch of control freaks and strive and live off [secrecy]. For years they have been with holding information that can change the world & put them into a panic of fear and confusion."[70] While governments have always deceived their citizens—Internet fact-checker Steve Benen documented 533 lies told by candidate Mitt Romney in the 2012 presidential campaign[71]—serious discussion of a "credibility gap" in the U.S. began during the Vietnam War, exploded during Watergate, and persists today on subjects as wide-ranging as climate change and suppressed cancer cures to use of unmanned killer drone aircraft.

Is Sasquatch included on the list of things concealed by our elected masters in the name of "national security"?

A Voice in the Wilderness

Few Sasquatch conspiracy theorists are more outspoken than Wayne Douglas Tarrant of Nevada, "Doug" to his friends and critics alike. A fascinating character, born in 1935, Tarrant has worked in law enforcement, toured as a musician with one incarnation of Bill Haley's Comets, and participated in filming Charles Pierce's classic 1971 mockumentary *The Legend of Boggy Creek,* all while pursuing Fortean studies across 46 U.S. states. Tarrant dates his first close encounter with Sasquatch from 1957. Fourteen years later, after members of the *Boggy Creek* film crew logged their own creature sighting, Tarrant hatched a plan that quickly went awry.

Since this was my chance to get me a Monster, and they were still called Monsters in 1971... I figured out a plan to get me a crew of hunters and go after the creature. Plan was to shoot him, freeze the body and get a lawyer and contact every anthropology department in every university and for the highest bid, offer a foot or hand or the whole cadaver for big bucks. That's when two government officials knocked on my door from the Governor's office. A "Cease & Desist" order. The Governor [Winthrop Rockefeller] said that Foukie was the state's pet for over 68 years in it's transit and never hurt anyone, and (with a wink wink) said that it will remain a MYTH.[72]

Later, during his decade as a deputy sheriff, Tarrant says he encountered the same official wall of silence on Sasquatch and similar "forbidden" subjects.

When I was a Deputy, if any call came in like for example: A Bigfoot was either hit by a car or shot dead, my watch commander had a special phone number that he would call, and a helicopter or truck would come out and pick up the carcass and it would just disappear. This is known as a government "cover-up." There would be no news media alerted. The event just didn't happen. Got the picture?? Same for a crashed UFO; it never happened.[73]

In his far-flung travels, Tarrant never missed an opportunity to seek new information on Sasquatch and other paranormal topics. One subject he interviewed was famed bow hunter Fred Bear (1902-88), who shared the following information.

> I talked to Mr. Fred Bear (Bear Archery) back in 1978, who had furnished the Smithsonian "Natural History" Museum with stuffed game animals from around the world, and he told me that the Smithsonian had logged in and had info on these creatures. About six cousins of the Yeti, Almas, Sasquatch, etc, at that time period, there was no mention of a tie with a Neanderthal mix. Mr. Bear had his own small museum down in Gainesville, Florida. He was 83 years old at the time I talked to him. Fred Bear told me the traveling Clyde Beatty Circus had acquired a YETI back in 1925, and was bringing it into the states after a world tour.

> The Smithsonian sent out three professors to examine and log any unknown animal import and found the Yeti to be too close to a human and advised the circus to let the being go before returning to the states. At that time, there was a monkey trial going on in Tennessee of the Scopes Trial verses [sic] a teacher who was teaching Evolution, and to expose a Yeti would blow the roofs off the churches ... It was covered up and the government was behind it. I think the Smithsonian's big brass keep the log books on unknown animals and the Government just backs them up. If the Sasquatch were proven to exist, there would be requests for the Government to set aside sanctioned lands/reservations so that the Beings would be protected to live out their lives in safety from hunters and prying spectators, etc.[74]

Was Bear strictly candid? Cryptozoologist Scott Marlowe writes that "Beatty didn't even found his circus until 1945. His expertise was in African animals and not Asian 'Yetis.' He did, however, appear in two B&W movies during his career that involved similar story lines."[75] No such Yeti-related features appear in Beatty's entry at the Internet Movie Database, but we know he left home to work with various circuses after graduating from high school in 1921, graduating to the role of "lion tamer" by 1923 and headlining for the Hagenbeck-Wallace Circus two years later.[75]

How high up does the conspiracy of silence extend, in Tarrant's opinion? In December 2011 he wrote: "I've known three CIA agents over the years and it's in the best interest of the mass (sheeple) to NOT KNOW of such things existing because of religious overtones and shutdown of prime forests. If proven to exist... and proven to be "humans"... the ACLU [American Civil Liberties Union] would demand human rights to sanction reservation land to protect the feral type humans. You remember what the spotted owl did for shutting down timber producing lands of the PNW [Pacific Northwest]. So we're dealing with a MYTH...... (wink wink)."[76]

Seven months later, in a post to the Forest Giants newsgroup, Tarrant described his own failed effort to promote a Sasquatch film based on purported actual events.

> Several years back, I pitched my screenplay to Hollywood of a TRUE STORY of two Bigfoot that killed four campers in the State of Georgia back in 1933. My great Uncle was involved and I got the full details of the situation as it unfolded... and as a cover up as well. I changed names and location and dates and embellished it into a story to make into a full length movie of interest. I was rejected by all agencies telling me to take out the "True Story" and resubmit it as FICTION and it would be considered. I just shelved the script...and when I learned that Dr. Ketchum was going to release her data, I was going to do a re-write of some parts of my script to detail the storyline to better sell the screenplay to parallel Ketchum's findings. Again...that all went "South" with the cover up.[77]

As for the Ketchum DNA report, Tarrants wrote: "Last year, Dr. Melba Ketchum and her copyrighted findings came out on CNN news and I caught that airing, and so did a few others only to be pulled off the air within the hour. Dr. Dmitri Bayanov was to air their findings on the YETI in tandem with Ketchum's report. Then it all went silent. Again... the 'Powers that Be' took charge and the cover up started again."[78]

State and federal authorities are not the only persons anxious to suppress proof of Sasquatch's existence, in Tarrant's view. Equally anxious are a group of believers—known to other researchers as "habituators"—who claim continuing long-term contact with manimals in North America. Tarrant writes: "When Janice Carter and Mary Green came out with their book '50 Years w/Bigfoot' [2002] it really upset the habituators. They formed a party to dispel the 'truth' and sent out rumors that it was a hoax. I don't know if Jan and Mary ever figured that one out but I do know that Loren Coleman certainly believed it was all a hoax; he helped spread that rumor around the 'net on his blog. Mary always complained that they had to put their own funds into publishing that book and lost money. The 'hoax' thing killed everything. The Government still keeps the creatures as a MYTH."[79]

Tarrant harbors no hope that renewed interest in Sasquatch over the past decade will produce any hard evidence. "Seems the only BF books that are being written and sold are only to the Bigfoot researchers themselves," he writes. "Matt the Moneymaker is getting his paychecks from his TV series of 'Finding Bigfoot.' Nothing is learned and nothing is new. Just entertainment for the viewers and money for the advertisers and TV production company. So make that buck when you can selling any books. As for the few of us who have had close up encounters with the elusive 'Beings,' we KNOW they are real and exist."[80]

Indeed, from Tarrant's viewpoint, even a documented kill or capture would be futile. He writes: "Regardless of what kind of luck you have with these methods, remember this. Nothing is going to be proved, or shot, or hit by a car without the 'powers that be' stepping in and covering it all up. That's been going on for decades... Nothing has changed except for a few observations and common sense and the anthropologist[s] are getting a leg up. Since the masses aren't anthropologist[s]... they will not learn a damn thing except what they are told. We are dealing with a type of HUMAN... some are ancient early Indians, some are feral and some are Neanderthals that are in pockets here and there and most have survived the HUMAN RACE that are us, and want nothing to do with us. I don't blame them. It's the curious juveniles that get sighted or almost hit by cars. The grownups know better and stay out of sight. So what does that leave us? Right back where we started. The Government covers it all up. Just MYTHS! A personal encounter is the only proof we will likely ever get, but to that individual only. That's as good as it will get. And believe me... up close and personal is enough!"[81]

Skofftics dismiss such observations as proof positive of paranoia—but are they mere agents of deception, champions of scientific orthodoxy lashing out in defense of their own careers as professional "skeptics"? Nothing is sure, beyond the fact that the debate persists.

Conclusion
Thus concludes our journey through the labyrinth of rumor and purported evidence of manimals gunned down, run down, captured alive, or found in sundry stages of decay. We end the search with no proof presently in hand, but will it ever be collected, Doug Tarrant's negative view notwithstanding?

Ray Crowe supports the cover-up scenario but still retains some cautious optimism. In 2011 he wrote: "If you should be so lucky as to find a bone or a body, GET PROOF! Dr. [Wolf-Henrich "Henner"] Fahrenbach says security is important. Move the specimen to somewhere neutral if you can, where land owners, police, etcetera, cannot make claims. Take many photographs, include your tape measure and do not take film to Wal-Mart. Put some body parts, a hand or foot, or even cut off the head (ugh) in baggies, and blood on a tissue. Even if rotten put any parts you can in your trunk; you can fumigate later. Do not leave without doing these important things first as the body would be gone when you returned, or if more than one of you stay and guard the body until scientists can be summoned."[1]

But who would respond? And can they be trusted? Robert Lindsay opines, "The state is not to be trusted one bit with Bigfoot evidence. Universities have a nasty habit of losing Bigfoot hard evidence, so we should not automatically turn evidence over to them."[2] What, then, should be done with any proof? Lindsay suggests:

> In order for science to make use of these bodies, the legal question regarding shooting a Bigfoot to death needs to be resolved somehow. Otherwise people who shoot and kill Bigfoots will continue to abandon them or bury them in the woods. Bigfoot organizations should establish procedures about what to do the next time a Bigfoot is shot and killed. Probably the best plan would be to say that the organization is willing to accept any Bigfoot shot dead, no questions asked. The person could then donate the body to the organization without fear of being prosecuted. It's doubtful that the government would go after the organization merely for holding a Bigfoot corpse. Anyone who shoots and kills a Bigfoot should try to protect the corpse and notify either Bigfoot organizations or prominent scientists such as Dr. Meldrum. Do not notify the authorities. If you do, you're likely to never see the body again.[3]

And, we might add, shun contacts with the media, since every kill or capture ballyhooed in headlines heretofore has proved to be a hoax, only increasing skepticism among scientists and the public at large. Lindsay takes the opposite view, seeing a paparazzi swarm as a kind of protection for Sasquatch hunters. "With the media on the scene taking pictures of the body," he writes, "it will be hard for the state to seal the area off and steal the body again."[4]

Concerning decomposed Sasquatch remains, Lindsay suggests, "If a Bigfoot burial is witnessed, notify Bigfoot researchers so they can excavate the gravesite. Try to take photos or movies of the burial. Possible Bigfoot graveyards should be excavated. Bodies, parts, bones, etc. should be turned over to Bigfoot researchers and not the state. Let the state conduct an armed raid to get them back."[5]

The image of raiders at large does not strike all Sasquatch hunters as rank paranoia. Hardly a day passes in the U.S. without news of another SWAT team strike against some target or other, frequently involving raids on wrong addresses, resulting in death or injury to completely innocent parties. In Congress, seemingly rational legislators warn of impending drone strikes against U.S. citizens on their native soil, sparked by vague accusations of terrorist sympathies. Nothing is inconceivable it seems, in the age of suspicion and increasingly bizarre "reality" entertainment.

We know only two things for certain. First, the private search for Sasquatch-type creatures continues, whatever its motives. And second, those who claim personal proof of Sasquatch's existence—the repeat eyewitnesses and the "habituators"—stubbornly refuse to offer any proof, either from an expressed desire to shield their "forest friends," or in the firm conviction that conspirators will simply make their evidence go up in smoke.

It is a stalemate, for the moment, and prospective jurors on the case can only voice the classic Scottish verdict of "not proven."

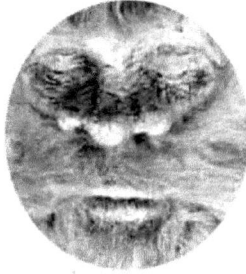

Notes

Introduction

1. Ray Crowe, "Where Are the Bigfoot Bones?" http://bigfootology.com/?p=406.
2. Robert Lindsay, "Why No Bigfoot Bones and Bodies?" http://robertlindsay.wordpress.com/2011/05/04/why-no-bigfoot-bones-and-bodies.

Chapter 1

1. Native American Bigfoot Names, http://www.bigfoot-lives.com/html/bigfoot_indian_names.html. Retrieved May 29, 2014.
2. Bigfoot Shootings, http://www.lawnflowersjerkyandbigfoots.com/Pages/BigfootShootings.aspx; Daniel Boone, http://en.wikipedia.org/wiki/Daniel_Boone. Both retrieved May 29, 2014.
3. Hugh H. Trotti, "Did Fiction Give Birth to Bigfoot?" *Skeptical Inquirer* 18 (Sept./Oct. 1994): 541-2.
4. Leonard Roberts, "Curious legend of the Kentucky mountains" *Western Folklore* 16 (1957): 48-51.
5. Randall Floyd, "Hunters told of swamp creature's attack," *Augusta* (GA) *Chronicle*, March 12, 2000.
6. Ibid.
7. Dr. Tuklo Nashoba, "The Legend of Sacred Baby Mountain," http://www.network54.com/Forum/61862/thread/1001354386/last-1027853702/THE+LEGEND+OF+SACRED+BABY+MOUNTAIN+By+Dr.+Tuklo+Nashoba. Retrieved May 30, 2014.
8. Personal communication with the author, May 30, 2014.
9. Janet and Colin Bord, *Bigfoot Casebook Updated* (Enumclaw, WA: Pine Winds Press, 2006), p. 220.
10. Nok-Noi Ricker, "Bigfoot in Maine? 10-foot-tall 'wild man' was killed in 1886, newspapers reported," *Bangor Daily News,* Oct. 27, 2013.
11. Ibid.
12. Bigfoots in Upper Alabama, http://www.lawnflowersjerkyandbigfoots.com/Pages/BigfootsinUpperAlabama.aspx. Retrieved May 30, 2014.

Chapter 2
1. International Bigfoot Society (IBS) reports #892 and 1261, http://www.mid-americanbigfoot.com/forums/viewtopic.php?f=167&t=3738&start=140. Retrieved June 16, 2014.
2. Sasquatch/Bigfoot, http://www.bcscc.ca/sasquatch.htm. Retrieved June 13, 2014.
3. Ibid.
4. Robert Lindsay, "Why No Bigfoot Bones and Bodies?"
5. Bill Oliver, "Bigfoot Tales From Kitimaat [*sic*] Village," http://www.ufobc.ca/Supernatural/Bigfoot/kittimatt.htm. Retrieved June 13, 2014.
6. Robert Lindsay, "Why Has No Hunter Ever Shot and Killed a Bigfoot?" http://robertlindsay.wordpress.com/2011/05/13/why-has-no-hunter-ever-shot-and-killed-a-bigfoot. Retrieved June 13, 2014.
7. Fred Beck, *I Fought the Apemen of Mount St. Helens, WA*, http://www.bigfootencounters.com/classics/beck.htm. Retrieved June 13, 2014.
8. Ibid.
9. Ibid.
10. Ibid.
11. Mark Sumerlin, "The Hoaxers," http://www.bigfootencounters.com/articles/fate2002.htm. Retrieved June 13, 2014.
12. John Green, *Sasquatch: The Apes Among Us* (North Vancouver, B.C.: Hancock House, 1978), p. 377.
13. Lindsay, "Why No Bigfoot Bones and Bodies?"; Lindsay, "Why Has No Hunter Ever Shot and Killed a Bigfoot?"
14. Sasquatch Tracker, http://webcache.googleusercontent.com/search?q=cache:9G9DfzcHX0IJ:sasquatchtracker.com/Database.html+&cd=4&hl=en&ct=clnk&gl=us&client=firefox-a. Retrieved June 13, 2014.
15. 18 Older Alaska Reports, http://www.bigfootencounters.com/sbs/oldalaska.htm. Retrieved June 13, 2014.
16. Green, p. 337.
17. Lindsay, "Why Has No Hunter Ever Shot and Killed a Bigfoot?"
18. Bigfoot Shootings, http://www.lawnflowersjerkyandbigfoots.com/Pages/BigfootShootings.aspx. Retrieved June 14, 2014.
19. Older Reports from the State of Missouri, http://www.bigfootencounters.com/sbs/oldermissouri.htm. Retrieved June 14, 2014.
20. John A. Keel, *The Complete Guide to Mysterious Beings* (New York: Doubleday, 1994), p. 120.
21. Lindsay, "Why Has No Hunter Ever Shot and Killed a Bigfoot?"
22. Bord, pp. 230-1.
23. All quotes in this section from BFRO Report #9552, http://www.bfro.net/gdb/show_report.asp?id=9552. Retrieved June 14, 2014.
24. Bigfoot Shootings.
25. Ibid.
26. Lindsay, "Why No Bigfoot Bones and Bodies?"
27. Green, p. 370.

Chapter 3

1. Bigfoot Shootings.
2. BFRO Report #8745, http://www.bfro.net/GDB/show_report.asp?id=8745. Retrieved June 15, 2014.
3. IBS Report #1551, http://www.mid-americabigfoot.com/forums/viewforum.php?f=167. Retrieved June 15, 2014.
4. Bigfoot Shootings.
5. Rick Berry, *Bigfoot on the East Coast* (Harrisonburg, VA: The author, 1993), p. 95.
6. Lindsay, "Why No Bigfoot Bones and Bodies?"
7. Crowe, "Where Are the Bigfoot Bones?"
8. Bord, p. 231.
9. Bigfoot Shootings.
10. California IBS Sighting Reports, http://www.mid-americabigfoot.com/forums/viewtopic.php?f=167&t=3984. Retrieved June 15, 2014.
11. Bigfoot Shootings.
12. Oregon IBS Sighting Reports, http://www.mid-americabigfoot.com/forums/viewtopic.php?f=167&t=5083. Retrieved June 15, 2014.
13. John Steele, "Can Bigfoot be Killed?" http://www.bigfootencounters.com/stories/jsteele.htm. Retrieved June 15, 2014.
14. Bigfoot Shootings.
15. GCBRO website, http://www.gcbro.com/TNov011.html. Retrieved June 15, 2014.
16. Bigfoot Shootings.
17. Oregonbigfoot.com File #00194, http://www.oregonbigfoot.com/report_detail.php?id=00194. Retrieved June 15, 2014.
18. Bord, p. 234.
19. Lindsay, "Why Has No Hunter Ever Shot and Killed a Bigfoot?"
20. Oregon IBS Sighting Reports, http://www.mid-americabigfoot.com/forums/viewtopic.php?f=167&t=5083&sid=49bb6aca4f85605a805b312e7aacdbb6. Retrieved June 16, 2014.
21. Ian Simmons, "The Abominable Showman," *Fortean Times* 83 (October 1995): 34-37.
22. Ivan Sanderson, "The Missing Link?" *Argosy* (May 1969): 23-31.
23. Frank Hansen, "I Killed The Ape-Man Creature Of Whiteface," *Saga* (July 1970): 8-11, 55-60.
24. Simmons, "The Abominable Showman."
25. William Jevning, "Minnesota Iceman," http://jevningresearch.blogspot.com/2011_12_01_archive.html. Retrieved June 16, 2014.
26. Andy Campbell, "Minnesota Iceman: Mysterious Frozen Creature From '60s Resurfaces At Museum," *Huffington Post,* June 28, 2013.
27. Bord, p. 236.
28. Bigfoot Shootings.
29. Lou Farish, "Fouke Monster Still Alive and Well," http://www.bfro.net/GDB/show_article.asp?id=258. Retrieved June 16, 2014.
30. Bigfoot Shootings.
31. Farish, "Fouke Monster Still Alive and Well."
32. Steele, "Can Bigfoot be Killed?"

33. Bigfoot Shootings.

34. Washington IBS Sightings Reports, http://www.mid-americabigfoot.com/forums/viewtopic.php?f=167&t=5547. Retrieved June 16, 2014.

35. Bigfoot Shootings.

36. IBS Report #194, http://www.mid-americabigfoot.com/forums/viewtopic.php?f=167&t=5824. Retrieved June 16, 2014.

37. Steele, "Can Bigfoot be Killed?"

38. .270 Winchester, http://en.wikipedia.org/wiki/.270_Winchester. Retrieved June 16, 2014.

39. Ithaca Mag-10, http://en.wikipedia.org/wiki/Ithaca_Mag-10. Retrieved June 16, 2014.

40. 18 Older Alaska Reports.

41. Steele, "Can Bigfoot be Killed?"

42. Oregonbigfoot.com file #00392, http://www.oregonbigfoot.com/report_detail.php?id=00392. Retrieved June 17, 2014.

43. Steele, "Can Bigfoot be Killed?"

44. GCBRO, http://www.gcbro.com/INcarroll001.html. Retrieved June 17, 2014.

45. Man Pumps Nine rounds into "Momo" from point blank range, http://www.bigfootencounters.com/stories/momo1969.htm. Retrieved June 17, 2014.

46. Sasquatch Tracker.

47. 18 Older Alaska Reports.

48. Bigfoot Shootings.

49. IBS Report #284, http://www.mid-americabigfoot.com/forums/viewtopic.php?f=167&t=4096. Retrieved June 17, 2014.

50. Washington IBS Sightings Reports.

51. 22nd Annual Ohio Bigfoot Conference, http://bigfootfieldreporter.com/wordpress/2010/page/45. Retrieved June 17, 2014.

52. Lindsay, "Why No Bigfoot Bones and Bodies?"

53. Oregon IBS Sighting Reports.

54. Bigfoot Shootings.

55. Shasta County, Wildwood, California—1960s , http://www.bigfootencounters.com/stories/shasta_countyCA.htm. Retrieved June 17, 2014.

Chapter 4.

1. Lindsay, "Why Has No Hunter Ever Shot and Killed a Bigfoot?"

2. Washington IBS Sightings Reports.

3. Raymond L. Wallace, http://en.wikipedia.org/wiki/Raymond_L._Wallace. Retrieved June 18, 2014.

4. Bigfoot Shootings.

5. Washington IBS Sightings Reports.

6. Bigfoot Shootings.

7. Bord, p. 258.

8. Bigfoot Shootings.

9. GCBRO, http://www.gcbro.com/FLcitrus0002.html. Retrieved June 18, 2014.

10. Washington IBS Sightings Reports.

11. Bigfoot Shootings.

12. GCBRO, http://www.gcbro.com/LAvernon0007.html. Retrieved June 18, 2014.

13. Bigfoot Shootings.
14. Berry, p. 86.
15. Bigfoot Shootings.
16. Personal communication with the author, June 18, 2014.
17. Bord, p. 264.
18. .30-30 Winchester, http://en.wikipedia.org/wiki/.30-30_Winchester. Retrieved June 18, 2014.
19. Bigfoot Shootings.
20. Ibid.
21. Berry, p. 80.
22. Reports in Rusk County, http://woodape.org/reports/report/county?county=Rusk&state=TX. Retrieved June 18, 2014.
23. Bigfoot Shootings.
24. Berry, p. 44.
25. Bigfoot Shootings.
26. Reports for Clinton County, KY, http://www.kentuckybigfoot.com/counties/clinton.htm. Retrieved June 18, 2014.
27. Bord, p. 274.
28. Bigfoot Shootings.
29. Berry, p. 102.
30. Fayette County Sightings, http://www.pabigfootsociety.com/sightingsfayette.html. Retrieved June 18, 2014.
31. Bigfoot Shootings.
32. Personal communication with the author, June 19, 2014.
33. Berry, p. 82.
34. Bord, p. 276.
35. Berry, p. 82.
36. Bord, p. 276.
37. Bigfoot Shootings.
38. State By State Sightings List, http://www.bigfootencounters.com/sbs/indiana_pa.htm. Retrieved June 19, 2014.
39. Bigfoot Shootings.
40. Berry, p. 102.
41. Bord, pp. 136-9.
42. Bigfoot Shootings.
43. Berry, p. 18.
44. Bigfoot Shootings.
45. Personal communication with the author, June 19, 2014.
46. Bord, p. 278.
47. Berry, p. 83.
48. Bord, p. 278.
49. Bigfoot Shootings.
50. Other Older Iowa Reports from the 1970s, http://www.bigfootencounters.com/sbs/olderiowa.htm. Retrieved June 19, 2014.
51. Bigfoot Shootings.

52. GCBRO, http://gcbro.com/FLcol001.html. Retrieved June 19, 2014.

53. Bigfoot Shootings.

54. Berry, p. 85.

55. Ibid., p. 84.

56. Bigfoot Shootings.

57. IBS Report #411, http://www.mid-americabigfoot.com/forums/viewtopic.php?f=167&t=4020. Retrieved June 19, 2014.

58. Older Oklahoma Reports, http://www.bigfootencounters.com/sbs/oldoklahoma.htm. Retrieved June 19, 2014.

59. Bigfoot Shootings.

60. Berry, p. 105.

61. Bigfoot Shootings.

62. IBS Report #1782, http://www.mid-americabigfoot.com/forums/viewtopic.php?f=167&t=3984&start=80. Retrieved June 19, 2014.

63. Bigfoot Shootings.

64. Personal communication with the author, June 19, 2014.

65. Bigfoot Shootings.

66. Berry, p. 50.

67. Lindsay, "Why Has No Hunter Ever Shot and Killed a Bigfoot?"

68. Bigfoot Shootings.

69. Berry, p. 88.

70. Bigfoot Shootings.

71. Oregon IBS Sighting Reports.

72. Bigfoot Shootings.

73. Dearborn County, Aurora, Indiana, http://www.bigfootencounters.com/sbs/aurora_IN.htm. Retrieved June 20, 2014.

74. Bigfoot Shootings.

75. Bigfoot Sighting Report #17, http://njbigfoot.org/show_report.php?17. Retrieved June 21, 2014.

76. GCBRO, http://www.gcbro.com/NJsussex0001.html. Retrieved June 21, 2014.

77. Berry, p. 91.

78. Ibid., p. 108.

79. Ibid., p. 92.

80. Bigfoot Shootings.

81. Fort Lewis, Washington, http://www.bigfootencounters.com/sbs/ftlewis1.htm. Retrieved June 21, 2014.

82. Bord, p. 302.

83. Ibid., p. 303.

84. Bigfoot Shootings.

85. Oregon IBS Sighting Reports.

86. Berry, p. 108.

87. Ibid., p. 109.

88. Bigfoot Shootings.

89. BFRO Media Article #195, http://www.bfro.net/GDB/show_article.asp?id=195. Retrieved June 22, 2014.

90. Bigfoot Shootings.
91. BFRO Report #24061, http://www.bfro.net/GDB/show_report.asp?id=24061. Retrieved June 22, 2014.
92. Sasquatch Tracker.
93. Alaska IBS Sighting Reports, http://www.mid-americabigfoot.com/forums/viewtopic.php?f=167&t=3738&start=220. Retrieved June 22, 2014.
94. Bigfoot Shootings.
95. IBS Report #2679, http://www.mid-americabigfoot.com/forums/viewforum.php?f=167. Retrieved June 23, 2014.
96. Bigfoot Shootings.
97. BFRO Media Article #195.
98. Bigfoot Shootings.
99. BFRO Report #2382, http://www.bfro.net/GDB/show_report.asp?id=2382. Retrieved June 23, 2014.
100. Bigfoot Shootings.
101. BFRO Report #792, http://www.bfro.net/GDB/show_report.asp?id=792. Retrieved June 23, 2014.
102. Bigfoot Shootings.
103. Berry, p. 20.
104. Bigfoot Shootings.
105. Berry, p. 57.
106. Reports in Cherokee County, http://woodape.org/reports/report/county?county=Cherokee&state=TX. Retrieved June 23, 2014.
107. Bigfoot Encounters, http://www.bigfootencounters.com/sbs/oglethorpeGA.htm. Retrieved June 23, 2014.
108. Oregon IBS Sighting Reports.
109. Bigfoot Shootings.
110. GCBRO, http://gcbro.com/OKcar001.htm. Retrieved June 23, 2014.
111. Bigfoot Shootings.
112. GCBRO, http://www.gcbro.com/OKpayn0002.html. Retrieved June 23, 2014.
113. Bigfoot Shootings.
114. GCBRO, http://www.gcbro.com/AZmaricopa0001.html. Retrieved June 23, 2014.
115. Bigfoot Shootings.
116. Virginia Bigfoot Research Organization, http://www.virginiabigfootresearch.org/read_sighting.asp?ID=52. Retrieved June 23, 2014.
117. Bigfoot Shootings.
118. Report #03080012, http://woodape.org/reports/report/detail/360. Retrieved June 23, 2014.
119. BFRO Report #742, http://www.bfro.net/GDB/show_report.asp?id=742. Retrieved June 23, 2014.

Chapter 5
1. Bigfoot Shootings.
2. BFRO Report #2394, http://www.bfro.net/GDB/show_report.asp?id=2394. Retrieved June 26, 2014.

3. Bigfoot Shootings.

4. GCBRO, http://www.gcbro.com/INcurr0001.htm. Retrieved June 26, 2014.

5. Bigfoot Shootings.

6. Report #02080013, http://woodape.org/reports/report/detail/399. Retrieved June 26, 2014.

7. Bigfoot Shootings.

8. GCBRO, http://www.gcbro.com/TNcumb003.html. Retrieved June 26, 2014.

9. Bigfoot Shootings.

10. Ibid.

11. IBS Report #164, http://www.mid-americabigfoot.com/forums/viewforum.php?f=167. Retrieved June 26, 2014.

12. IBS Report #2402, http://www.mid-americabigfoot.com/forums/viewforum.php?f=167. Retrieved June 26, 2014.

13. IBS Report #2539, http://www.mid-americabigfoot.com/forums/viewforum.php?f=167. Retrieved June 26, 2014.

14. Ibid.

15. Ibid.

16. Ibid.

17. Loren Coleman, "Bugs: Where Are You?" http://cryptomundo.com/cryptozoo-news/bugs-morgan. Retrieved June 26, 2014.

18. Loren Coleman, "Bugs Bigfoot Legend Begun By Radio Freakazoid," http://cryptomundo.com/cryptozoo-news/bugs-freakazoid. Retrieved June 26, 2014.

19. "Bugs," bigfoot and Ed Hale, http://mountainsageblog.com/2009/01/05/bugs-bigfoot-and-ed-hale. Retrieved June 26, 2014.

20. Bigfoot Shootings.

22. Alabama Bigfoot Society, http://alabamabigfootsociety.com. Retrieved June 26, 2014.

22. Oregonbigfoot.com file #00821, http://www.oregonbigfoot.com/report_detail.php?id=00821. Retrieved, June 26, 2014.

23. Bigfoot Shootings.

24. Sasquatch Information Society, http://www.bigfootinfo.org/sightings/sightings-database. Retrieved June 30, 2014.

25. Bigfoot Shootings.

26. GCBRO, http://www.gcbro.com/TNpickett001.htm. Retrieved June 30, 2014.

27. Ibid.

28. Report #01020052, http://woodape.org/reports/report/detail/186. Retrieved June 30, 2014.

29. Bigfoot Shootings.

30. GCBRO, http://www.gcbro.com/TNhawkins001.html. Retrieved June 30, 2014.

31. GCBRO, http://www.gcbro.com/OKmur0001.htm. Retrieved June 30, 2014.

32. IBS Report #3537, http://www.mid-americabigfoot.com/forums/viewforum.php?f=167. Retrieved June 30, 2014.

33. Bigfoot Shootings.

34. BFRO Report #1299, http://www.bfro.net/GDB/show_report.asp?id=1299. Retrieved June 30, 2014.

35. Lindsay, "Why Has No Hunter Ever Shot and Killed a Bigfoot?"

36. The "Siege" of Honobia, http://www.bfro.net/avevid/ouachita/siege-at-honobia.asp. Retrieved June 30, 2014.

37. Ibid.
38. Ibid.
39. Ibid.
40. Letter to the author from FBI headquarters, dated July 3, 2014.
41. Bigfoot Shootings.
42. GCBRO, http://www.gcbro.com/ALrussel001.html. Retrieved July 4, 2014.
43. Bigfoot Shootings.
44. GCBRO, http://www.gcbro.com/inorange001.html. Retrieved July 4, 2014.
45. Bigfoot Shootings.
46. GCBRO, http://gcbro.com/aldb1.htm. Retrieved July 4, 2014.
47. Bigfoot Shootings.
48. Report #02080016, http://woodape.org/reports/report/detail/408. Retrieved July 4, 2014.
49. Bigfoot Shootings.
50. Alabamabigfoot.com, http://www.alabamabigfoot.com/cgi-sys/suspendedpage.cgi. Checked on July 4, 2014.
51. Bigfoot Shootings.
52. Walker County, Texas, http://www.bfro.net/GDB/show_county_reports.asp?state=tx&county=Walker. Retrieved July 4, 2014.
53. Report #01030015, http://woodape.org/reports/report/detail/232. Retrieved July 4, 2014.
54. Bigfoot Shootings.
55. Guy Barnes, "Bigfoot Has Long, Lively Local History," http://www.bigfootencounters.com/articles/decatur.htm. Retrieved July 4, 2014.
56. Bigfoot Shootings.
57. Westmoreland County Sightings, http://www.pabigfootsociety.com/sightingswestmoreland.html. Retrieved July 4, 2014.
58. Bigfoot Shootings.
59. Oregonbigfoot.com file #01017, http://www.oregonbigfoot.com/report_detail.php?id=01017. Retrieved July 4, 2014.
60. Bigfoot Shootings.
61. Texas IBS Sighting Reports, http://www.mid-americabigfoot.com/forums/viewtopic.php?f=167&t=5108. Retrieved July 4, 2014.
62. Oregonbigfoot.com file #05498, http://www.oregonbigfoot.com/report_detail.php?id=05498.
Retrieved July 4, 2014.
63. Guy Edwards, "Thom Powell Week: On the heels of the PG film Anniversary," http://www.bigfootlunchclub.com/2010/10/thom-powell-week-on-heels-of-pg-film.html. Retrieved July 4, 2014.
64. Loren Coleman, "Bigfoot Massacre Theorist, John Green & Coverup," http://cryptomundo.com/cryptozoo-news/davis-back. Retrieved July 20, 2014.
65. Guy Edwards, "M. K. Davis: Video Proof Bigfoot was Shot in Thigh," http://www.bigfootlunchclub.com/2010/09/mkdavis-video-proof-bigfoot-was-shot-in.html. Retrieved July 20, 2014.
66. Bill Miller, "The Massacre at Bluff Creek…just when you think you've heard it all!" www.sasquatch-bc.com/massacre.html. Retrieved July 20, 2014.
67. Ibid.

68. Steven Streufert, "Blue Creek Mountain and Bluff Creek Bigfoot Timeline," http://bigfootbooksblog.blogspot.com/2009/12/blue-creek-mountain-and-bluff-creek.html. Retrieved July 20, 2014.
69. William Jevning, "Minnesota Iceman," http://jevningresearch.blogspot.com/2011_12_01_archive.html. Retrieved July 20, 2014.
70. Ibid.
71. Loren Coleman, "Ultimate GA Bigfoot Hoax Timeline: 2008," http://cryptomundo.com/cryptozoo-news/hoax-tl-08. Retrieved July 20, 2014.
72. Ibid.
73. Ibid.
74. Loren Coleman, "Ultimate GA Bigfoot Hoax Timeline," http://cryptomundo.com/cryptozoo-news/ultimate-timeline. Retrieved July 20, 2014.
75. Ibid.
76. Ibid.
77. Ibid.
78. Ibid.
79. Coleman, "Ultimate GA Bigfoot Hoax Timeline: 2008."
80. Ibid.
81. Ibid.
82. Ibid.
83. "Bigfoot hoaxers say it was just 'a big joke'," CNN, Aug. 21, 2008.
84. Guy Edwards, "Bigfoot Hoaxer, Rick Dyer, Arrested for eBay Fraud," http://www.bigfootlunchclub.com/2011/01/bigfoot-hoaxer-rick-dyer-arrested-for.html. Retrieved July 23, 2014.
85. Bigfoot Killed in San Antonio? http://www.snopes.com/photos/supernatural/bigfoot2014.asp. Retrieved July 23, 2014.
86. Ibid.
87. Lee Speigel, "Bigfoot 'Killer' Rick Dyer: 'It's Really Easy To Trick People'," *Huffington Post,* February 27, 2014.
88. Shawn Evidence, "Sierra Kills Bigfoot shooter cracks under pressure, reveals location and everything under the sun," http://bigfootevidence.blogspot.com/2011/10/sierra-kills-bigfoot-shooter-cracks.html. Retrieved July 23, 2014.
89. Ibid.
90. Shawn Evidence, "Sierra Kills driver speaks up, gives exact date of Bigfoot shooting. Describes what happened," http://bigfootevidence.blogspot.com/2011/11/sierra-kills-driver-speaks-up-gives.html. Retrieved July 23, 2014.
91. Evidence, "Sierra Kills Bigfoot shooter cracks."
92. The DNA Results of the Justin Smeja Kill—(not a bigfoot), http://bigfootfieldreporter.com/wordpress/2013/03/01/the-dna-results-of-the-justin-smeja-kill-not-a-bigfoot. Retrieved Aug. 3, 2014.
93. Trent DNA Report, http://www.sierrasiteproject.com/p/trent-university-dna-results.html. Retrieved Aug. 3, 2014.
94. Smeja Family History, http://www.ancestry.com/name-origin?surname=smeja. Retrieved Aug. 3, 2014.
95. Technical Examination Report, https://docs.google.com/file/d/0B-

ogATODxdKVLVlfR080NTBaQU0/edit?pli=1. Retrieved Aug. 3, 2014.

96. Sean Evidence, "Smeja Hoaxed Movie Sighting of Bigfoot," http://bigfootevidence101.blogspot.com/2013/09/smeja-hoaxed-movie-sighting-of-bigfoot.html. Retrieved Aug. 3, 2014.

97. DEAD BIGFOOT: A True Story, http://www.bigfootcrossroads.com/2013/11/dead-bigfoot-true-story.html. Retrieved Aug. 3, 2014.

98. DEAD BIGFOOT: A True Story, http://www.deadbigfoot.com. Retrieved Aug. 3, 2014.

99. Bigfoot Shootings.

100. IBS Report #1274, http://www.mid-americabigfoot.com/forums/viewtopic.php?f=167&t=4022. Retrieved Aug. 3, 2014.

101. Bigfoot Shootings.

102. Oregonbigfoot.com File #00743, http://www.oregonbigfoot.com/report_detail.php?id=00743. Retrieved Aug. 3, 2014.

103. Bigfoot Shootings.

104. Oklahoma IBS Sighting Reports, http://www.mid-americabigfoot.com/forums/viewtopic.php?f=167&t=4106. Retrieved Aug. 3, 2014.

105. Bigfoot Shootings.

106. Washington IBS Sightings Reports, http://www.mid-americabigfoot.com/forums/viewtopic.php?f=167&t=5547. Retrieved Aug. 3, 2014.

107. Bigfoot Shootings.

108. Harrison County, West Virginia, http://www.bigfootencounters.com/sbs/harrisoncounty.htm. Retrieved Aug. 3, 2014.

109. Ibid.

110. Lindsay, "Why No Bigfoot Bones and Bodies?"

111. IBS Reports #145, 2248 and 3342, http://www.mid-americanbigfoot.com/forums/viewtopic.php?f=167&t=3738&start=120. Retrieved Aug. 4, 2014.

112. IBS Report #3342.

113. Lindsay, "Why No Bigfoot Bones and Bodies?"

114. Crowe, "Where Are the Bigfoot Bones?"

115. Lindsay, "Why No Bigfoot Bones and Bodies?"

116. Crowe, "Where Are the Bigfoot Bones?"

117. Kyle Mizokami, "Datus Perry," http://www.bigfootencounters.com/stories/datus-perry.htm. Retrieved Aug. 4, 2014.

118. Crowe, "Where Are the Bigfoot Bones?"

119. Ibid.

120. Ibid.

121. Ibid.

122. Lindsay, "Why No Bigfoot Bones and Bodies?"

123. Crowe, "Where Are the Bigfoot Bones?"

124. Lindsay, "Why No Bigfoot Bones and Bodies?"

125. Crowe, "Where Are the Bigfoot Bones?"

126. Rob Davis, "Professor Bob's Blieve [*sic*] It or Not!" http://www.dmagazine.com/publications/d-magazine/1993/june/professor-bobs-blieve-it-or-not. Retrieved Aug. 4, 2014.

127. Ibid.

128. Crowe, "Where Are the Bigfoot Bones?"

129. Washington IBS Sightings Reports.

Chapter 6.
1. "Man hit by car twice in apparent 'Bigfoot' hoax," Associated Press, Aug. 258, 2012.
2. Where Are All the Bigfoot Roadkills? An Updated Analysis Using Mammal Roadkill Data, http://thoughtsonscienceandpseudoscience.blogspot.com/2012/12/an-updated-analysis-of-animal-roadkill.html. Retrieved Aug. 5, 2014.
3. Ibid.
4. Bord, pp. 227-315.
5. Pedestrian and Bicycle Information Center, http://www.pedbikeinfo.org/data/faq_details.cfm?id=31. Retrieved Aug. 5, 2014.
6. How come there's no bigfoot roadkill? http://www.bfro.net/gdb/show_FAQ.asp?id=410. Retrieved Aug. 5, 2014.
7. Lindsay, "Why No Bigfoot Bones and Bodies?"
8. Train Hits Sasquatch, http://daruc.pagesperso-orange.fr/white4.htm. Retrieved Aug. 5, 2014.
9. Lindsay, "Why No Bigfoot Bones and Bodies?"
10. Crowe, "Where Are the Bigfoot Bones?"
11. IBS Report #1458, http://www.mid-americabigfoot.com/forums/viewforum.php?f=167. Retrieved Aug. 5, 2014.
12. Crowe, "Where Are the Bigfoot Bones?"
13. Green, p. 409.
14. Ibid., p. 410.
15. Ibid., p. 278; Bord, p. 276.
16. Green, p. 278.
17. Bord, p. 280.
18. Green, p. 227.
19. Bord, p. 281.
20. Washington IBS Sightings Reports.
21. Bord, p. 168.
22. Ibid.
23. IBS Report #1298, http://www.mid-americabigfoot.com/forums/viewforum.php?f=167. Retrieved Aug. 7, 2014.
24. Personal communications with the author, April 18-21, 2014.
25. Oregonbigfoot.com File #01051, http://www.oregonbigfoot.com/report_detail.php?id=01051. Retrieved Aug. 7, 2014.
26. IBS Reports #2494 and 3486, http://www.mid-americabigfoot.com/forums/viewtopic.php?f=167&t=5547. Retrieved Aug. 17, 2014.
27. Crowe, "Where Are the Bigfoot Bones?"
28. IBS Report #1214, http://www.mid-americabigfoot.com/forums/viewforum.php?f=167. Retrieved Aug. 15, 2014.
29. Ibid.
30. Ibid.

Chapter 7.

1. Sally Lou Hock, "Bigfoot burials or graveyards," http://www.howwwl.com/t/Bigfoot/6736. Retrieved Aug. 8, 2014.

2. Frequently-Asked Questions, http://www.bfro.net/gdb/show_FAQ.asp?id=932. Retrieved Aug. 8, 2014.

3. Crowe, "Where Are the Bigfoot Bones?"

4. Ibid.

5. Lindsay, "Why No Bigfoot Bones and Bodies?"

6. Green, p. 368.

7. Crowe, "Where Are the Bigfoot Bones?"

8. Ibid.

9. Lindsay, "Why No Bigfoot Bones and Bodies?"

10. Ray Crowe, "Where Are the Bigfoot Bones Hidden?" http://sasquatch-pg.net/Burry%20Their%20Dead.htm. Retrieved April 9, 2014; no longer active.

11. Ibid.

12. Lindsay, "Why No Bigfoot Bones and Bodies?"

13. John A. Bindernagel, "Sasquatches in Our Woods," http://woodape.org/index.php/about-bigfoot/articles/89-sasquatches-in-our-woods. Retrieved Aug. 8, 2014.

14. Craig Woolheater, "John Green on Glen Thomas and William Roe," http://cryptomundo.com/bigfoot-report/john-green-on-thomas-and-roe. Retrieved Aug. 8, 2014.

15. James A. Hewkin, "Sasquatch Investigations in the Pacific Northwest," http://www.bigfootencounters.com/biology/hewkin92.htm. Retrieved Aug. 8, 2014.

16. Crowe, "Where Are the Bigfoot Bones?"

17. Crowe, "Where Are the Bigfoot Bones Hidden?"

18. Norma Gaffron, *Bigfoot* (Farmington Hills, MI: Greenhaven Press, 1988), p. 88.

19. Robert and Francis Guenette, *Bigfoot: The Mysterious Monster* (Los Angeles: Schick Sun Classic Pictures, 1975), p. 102.

20. Crowe, "Where Are the Bigfoot Bones Hidden?"

21. Lindsay, "Why No Bigfoot Bones and Bodies?"

22. Ibid.

23. IBS Report #1497, http://www.mid-americabigfoot.com/forums/viewforum.php?f=167. Retrieved Aug. 9, 2014.

24. Lindsay, "Why No Bigfoot Bones and Bodies?"

25. Crowe, "Where Are the Bigfoot Bones Hidden?"

26. Lindsay, "Why No Bigfoot Bones and Bodies?"

27. Crowe, "Where Are the Bigfoot Bones?"

28. Kirsten Stanley, "Bigfoot Burial Ground Identified In Ohio?" *Portsmouth Daily Times,* May 25, 1999.

29. Lindsay, "Why No Bigfoot Bones and Bodies?"

30. BFRO Report #5176, http://www.bfro.net/gdb/show_report.asp?id=5176. Retrieved Aug. 9, 2014.

31. Watch: Park Ranger Witnesses Bigfoot Burial, http://bigfootevidence.blogspot.com/2014/04/watch-park-ranger-witnesses-bigfoot.html. Retrieved Aug. 9, 2014.

32. Washington IBS Sightings Reports.

33. Lindsay, "Why No Bigfoot Bones and Bodies?"
34. Washington IBS Sightings Reports.
35. Crowe, "Where Are the Bigfoot Bones?"
36. Lindsay, "Why No Bigfoot Bones and Bodies?"
37. Crowe, "Where Are the Bigfoot Bones?"
38. Lindsay, "Why No Bigfoot Bones and Bodies?"
39. Crowe, "Where Are the Bigfoot Bones?"
40. Sasquatch Tracker, http://webcache.googleusercontent.com/search?
q=cache:9G9DfzcHX0IJ:sasquatchtracker.com/
Database.html+&cd=4&hl=en&ct=clnk&gl=us&client=firefox-a. Retrieved Aug. 10, 2014.
41. IBS Reports #145, 2248 and 3342, http://www.mid-americanbigfoot.com/forums/
viewtopic.php?f=167&t=3738&start=120. Retrieved Aug. 10, 2014.
42. Ibid.

Chapter 8

1. Lindsay, "Why No Bigfoot Bones and Bodies?"
2. Grover S. Krantz, *Big Footprints: A Scientific Inquiry into the Reality of Sasquatch* (Boulder, CO: Johnson Books, 1992), p. 130.
3. The Breitwinner Caves, http://greaterancestors.com/the-bretweiner-caves. Retrieved Aug. 10, 2014.
4. Northumberland, England giants, http://greaterancestors.com/northumberland-england-giants. Retrieved Aug. 10, 2014.
5. Salisbury 9 feet 4 inches, http://greaterancestors.com/salisbury-9-feet-4-inches. Retrieved Aug. 10, 2014.
6.Tallest Man, http://web.archive.org/web/20100319004913/http://www.guinnessworldrecords.com/records/human_body/extreme_bodies/tallest_man.aspx. Retrieved Aug. 10, 2014.
7. Joseph Comstock, *The Tongue of Time, and Star of the States: A System of Human Nature, with the Phenomena of the Heavens and Earth* (New York: The author, 1838), pp. 86-7.
8. "Skeleton of a Giant," *Oswego Commercial Times,* Aug. 8, 1851.
9. "Two Very Tall Skeletons," *New York Times,* Aug. 10, 1880.
10. Henry R. Schoolcraft, "Observations respecting the Grave Creek Mound," *Transactions of the American Ethnological Society* 1 (1845): 368-420.
11. Dave Cain, "Giants in our Midst? Tall Skeletons Reported Found in Marion County, WV," http://www.bibliotecapleyades.net/gigantes/MarionCounty.html. Retrieved Aug. 11, 2014.
12. "The Grave of a Giant," *Towanda Daily Review,* Oct. 25, 1897.
13. USA Place Names, http://www.placenames.com/us. Retrieved Aug. 11, 2014.
14. Lindsay, "Why No Bigfoot Bones and Bodies?"
15. "Skeleton of a Giant Found," *New York Times*, Nov. 21, 1856.
16. IBS Report #905, http://www.mid-americabigfoot.com/forums/viewforum.php?f=167. Retrieved Aug. 11, 2014.
17. Crowe, "Where Are the Bigfoot Bones?"
18. "Reported Discovery of a Huge Skeleton," *New York Times,* Dec. 25, 1868.
19. "A Remarkable Sight," *Daily Telegraph,* Aug. 23, 1871.
20. "More Big Indians Found in Virginia," *New York Times,* Sept. 8, 1871.

21. "The Bones of a Giant Found," *New York Times,* May 25, 1882.

22. "A Giant's Remains in a Mound," *New York Times,* Nov. 20, 1883.

23. "Skeletons Seven Feet Long," *New York Times,* May 5, 1885.

24. Lindsay, "Why No Bigfoot Bones and Bodies?"

25. Ibid.

26. Calhoun County A1Archives News, http://files.usgwarchives.net/al/calhoun/newspapers/newspape1131gnw.txt. Retrieved Aug. 11, 2014.

27. "A Race of Indian Giants," *New York Times,* Feb. 9, 1890.

28. "A Pre-Historic Giant," *The Popular Science News & Boston Journal of Chemistry* 24 (August 1890): 1.

29. Untitled article, *Coconino Sun*, Aug. 22, 1895.

30. "Cave in Mexico Gives Up the Bones of an Ancient Race," *New York Times,* May 4, 1908.

31. *New York Herald-Tribune,* June 21, 1925.

32. Nayarit, Mexico, Giant Skeletons, http://greaterancestors.com/nayarit-mexico-giant-skeletons. Retrieved Aug. 12, 2014.

33. "Blond Giant Remains Found," *Kentucky New Era,* Jan. 2, 1951.

34. "Wisconsin Mound Opened," *New York Times,* Dec. 20, 1897.

35. "Skeletons of Giants in Alaska," *San Francisco Call*, Nov. 18, 1900.

36. "Giant Skeletons Found, *New York Times,* Feb. 11, 1902.

37. "Find Giant Indians' Bones," *New York Times,* Sept. 7, 1904.

38. "Strange Skeletons Found," *New York Times,* May 4, 1912.

39. "Giants' Bones in Mound," *New York Times,* July 14, 1916.

40. Lindsay, "Why No Bigfoot Bones and Bodies?"

41. "New Link in Man History Is Found on West Coast," *Decatur Weekly Republican,* April 9, 1923.

42. "Gigantic Skeletons Located in Ontario," *Sarasota Herald,* Oct. 24, 1934.

43. "Giant Skeletons Are Discovered," *Milwaukee Sentinel*, Oct. 21, 1934.

44. "Beach Giant's Skull Unearthed By WPA Workers Near Victoria," *San Antonio Express*, Jan. 7, 1940.

45. "Skeleton Found of Super-Indian," *The Leader-Post* (Victoria, BC), May 28, 1943.

46. Lindsay, "Why No Bigfoot Bones and Bodies?"

47. Ibid.

48. Ibid.

49. Reports from Wolfe County, KY, http://www.kentuckybigfoot.com/reports.htm. Retrieved Aug. 12, 2014.

50. Lindsay, "Why No Bigfoot Bones and Bodies?"

51. Matt Moneymaker, "Buried Treasure: The Minaret Skull," http://www.bfro.net/REF/THEORIES/MJM/minaret.htm. Retrieved Aug. 12, 2014.

52. Ibid.

53. Ibid.

54. Washington IBS Sightings Reports.

55. Green, p. 373.

56. Ibid., pp. 373-4.

57. Ibid., p. 374.

58. Crowe, "Where Are the Bigfoot Bones?"

59. Lindsay, "Why No Bigfoot Bones and Bodies?"

60. Ivan T. Sanderson, *Abominable Snowmen: Legend Come to* Life (Philadelphia: Chilton, 1961), pp. 36-7.

61. Lindsay, "Why No Bigfoot Bones and Bodies?"

62. IBS Report #1506, http://www.mid-americabigfoot.com/forums/viewforum.php?f=167. Retrieved Aug. 14, 2014.

63. Lindsay, "Why No Bigfoot Bones and Bodies?"

64. IBS Report #1505, http://www.mid-americabigfoot.com/forums/viewforum.php?f=167. Retrieved Aug. 14, 2014.

65. IBS Report #1570, http://www.mid-americabigfoot.com/forums/viewforum.php?f=167. Retrieved Aug. 14, 2014.

66. Ray Crowe, "Early Man as a Model for Bigfoot," http://www.bigfootencounters.com/biology/earlyman.htm. Retrieved Aug. 15, 2014.

67. Lindsay, "Why No Bigfoot Bones and Bodies?"

68. Momo: Missouri's Legendary Monster, http://www.phantomsandmonsters.com/2011/09/momo-missouris-legendary-monster.html. Retrieved Aug. 15, 2014.

69. Lindsay, "Why No Bigfoot Bones and Bodies?"

70. IBS Report #3670, http://www.mid-americabigfoot.com/forums/viewtopic.php?f=167&t=3738. Retrieved Aug. 15, 2014.

71. Geo-Hack—Revillagigedo Island, http://tools.wmflabs.org/geohack/geohack.php?pagename=Revillagigedo_Island&params=55_38_03_N_131_17_51_W_scale:1000000. Retrieved Aug. 15, 2014.

72. Lindsay, "Why No Bigfoot Bones and Bodies?"

73. Andy Campbell, "Bigfoot DNA Tests: Melba Ketchum's Research Results Are Bogus, Claims Houston Chronicle Report," *Huffington Post,* July 3, 2013.

74. Lindsay, "Why No Bigfoot Bones and Bodies?"

75. Crowe, "Where Are the Bigfoot Bones?"

76. Oregon IBS Sighting Reports.

77. Lindsay, "Why No Bigfoot Bones and Bodies?"

78. Montana IBS Sighting Reports, http://www.mid-americabigfoot.com/forums/viewtopic.php?f=167&t=4075. Retrieved Aug. 15, 2014.

79. Kathryn Showen, *Great Falls Tribune* news assistant, personal communication dated Aug. 15, 2014.

80. Lindsay, "Why No Bigfoot Bones and Bodies?"

81. IBS Report #4122, http://www.mid-americabigfoot.com/forums/viewtopic.php?f=167&t=4075. Retrieved Aug. 15, 2014/

82. Crowe, "Where Are the Bigfoot Bones?"

83. Ibid.

84. IBS Report #3496, http://www.mid-americabigfoot.com/forums/viewtopic.php?f=167&t=3738&start=200. Retrieved Aug. 15, 2014.

85. Lindsay, "Why No Bigfoot Bones and Bodies?"

86. GCBRO, http://www.gcbro.com/OHcosh001.html. Retrieved Aug. 15, 2014.

87. Lindsay, "Why No Bigfoot Bones and Bodies?"

88. Crowe, "Where Are the Bigfoot Bones Hidden?"

89. Crowe, "Where Are the Bigfoot Bones?"

90. Shuswap Lake, http://en.wikipedia.org/wiki/Shuswap_Lake. Retrieved Aug. 15, 2014.

91. Alan Landsburg, *In Search of Myths and Monsters* (New York: Bantam, 1977), p. 127.

92. Crowe, "Where Are the Bigfoot Bones?"

93. Lindsay, "Why No Bigfoot Bones and Bodies?"

94. IBS Report #1536, http://www.mid-americabigfoot.com/forums/viewforum.php?f=167. Retrieved Aug. 16, 2014.

95. Ibid.

96. Lindsay, "Why No Bigfoot Bones and Bodies?"

97. Ibid.

98. Crowe, "Where Are the Bigfoot Bones?"; Oregon IBS Bigfoot Sightings.

99. Washington IBS Sightings Reports.

100. Ibid.

101. IBS Report #297, http://www.mid-americabigfoot.com/forums/viewtopic.php?f=167&t=5083&start=580. Retrieved Aug. 16, 2014.

102. IBS Report #1702, http://www.mid-americabigfoot.com/forums/viewtopic.php?f=167&t=5083&start=200. Retrieved Aug. 16, 2014.

Chapter 9

1. Lindsay, "Why No Bigfoot Bones and Bodies?"

2. John Peabody Harrington, http://www.keepersofindigenousways.org/id12.html. Retrieved Aug. 16, 2014.

3. Great Story: Wild Man Captured with Two Cubs 1839, http://www.bigfootencounters.com/stories/captured_wildman1839.htm. Retrieved Aug. 16, 2014.

4. Historic Sasquatch Sightings: Maine, http://www.bigfootencounters.com/sbs/campingout.htm. Retrieved Aug. 16, 2014.

5. Bord, p. 219.

6. "A Wild Man in Waldoboro," *Pioneer and Democrat,* May 12, 1855, https://www.sos.wa.gov/legacy/images/newspapers/SL_dir_olympiapiondemo/pdf/SL_dir_olympiapiondemo_05121855.pdf. Retrieved Aug. 16, 2014.

7. Otto Ernest Raymond, *Ozark Country* (New York: Duell, Sloan and Pearce, 1941), pp. 313-14.

8. Reprinted in Chad Arment, *The Historical Bigfoot* (Landisville, PA: Coachwhip Publications, 2006), p. 304.

9. Ibid., p. 305.

10. Ibid., pp. 306-8.

11. Ibid., 56-8.

12. Ibid., p. 58.

13. Ibid., p. 59.

14. Lindsay, "Why Has No Hunter Ever Shot and Killed a Bigfoot?"

15. W. H. Whimple, "The Wild Family," *Republic County Pilot,* Sept. 9, 1886.

16. Lindsay, "Why No Bigfoot Bones and Bodies?"

17. List of battles fought in Kansas, http://en.wikipedia.org/wiki/List_of_battles_fought_in_Kansas. Retrieved Aug. 17, 2014.

18. Lindsay, "Why No Bigfoot Bones and Bodies?"

19. 18 Older Alaska Reports, http://www.bigfootencounters.com/sbs/oldalaska.htm. Retrieved

Aug. 17, 2014.

20. MABRC, http://www.mid-americabigfoot.com/index.php/join-the-mabrc. Retrieved Aug. 17, 2014.

21. New Captured Bigfoot Claim....Speciman [*sic*] Being Examined, http://beforeitsnews.com/paranormal/2012/12/new-captured-bigfoot-claim-speciman-being-examined-2446354.html. Retrieved Aug. 17, 2014.

22. Ibid.

23. Ibid.

24. Ibid.

25. Ibid.

26. Rob Gaudet, "QUantra Bigfoot Capture Story, Separating Fact from Fiction," http://squatchunlimited.org/profiles/blogs/the-capture-of-bigfoot-by-quantra-separating-fact-from-fiction. Retrieved Aug. 17, 2014.

27. Melissa Adair, "Bigfoot Chicks interview with Darren Lee, the founder of MABRC," http://bigfootchicks.blogspot.com/2012/12/breaking-original-6-have-bigfoot-in-box.html. Retrieved Aug. 17, 2014.

28. Shawn Evidence, "The Bigfoot Report—Bigfoot News #11—The Quantra Group Live Capture Update," https://www.youtube.com/watch?v=Iz85X2Qonb0. RetNamed," https://squatchdetective.wordpress.com/tag/team-quantra. Retrieved Aug. 18, 2014.rieved Aug. 17, 2014.

29. Steve Kulls, "Breaking News—Bridging Group

30. Darren Lee, "Darkwing's Public Statement about the Quantra Affair," http://bigfootfieldreporter.com/wordpress/2013/01/14/darkwings-public-statement-about-the-quantra-affair. Retrieved Aug. 18, 2014.

31. Personal communication with the author, Aug. 18, 2014.

Chapter 10
1. Almas (cryptozoology), http://en.wikipedia.org/wiki/Almas_%28cryptozoology%29. Retrieved Aug. 18, 2014.

2. Was Russian "Bigfoot" actually an African slave? http://www.channel4.com/info/press/news/was-russian-bigfoot-actually-an-african-slave. Retrieved Aug. 19, 2014.

3. Igor Bourtsev, "A Skeleton Still Buried and a Skull Unearthed: The Story of Zana," http://www.bigfootencounters.com/articles/zana.htm. Retrieved Aug. 19, 2014.

4. Almas (cryptozoology).

5. Sharon Hill, "The story of 'Zana,' wild woman, has been solved through DNA analysis," http://doubtfulnews.com/2013/11/the-story-of-zana-wild-woman-has-been-solved-through-dna-analysis. Retrieved Aug. 19, 2014.

6. Brian Dunning, "De Loys' Ape," http://skeptoid.com/episodes/4302. Retrieved Aug. 19, 2014.

7. Ibid.

8. Ibid.

9. Ibid.

10. George Montandon, http://fr.wikipedia.org/wiki/George_Montandon. Retrieved Aug. 19, 2014.

11. Dunning, "De Loys' Ape."

12. Michel Raynal, "De Loys' Well-Known Prank," http://cryptomundo.com/cryptozoo-news/raynal-de-loys. Retrieved Aug. 19, 2014.

13. Ibid.

14. Dunning, "De Loys' Ape."

15. The Hairy Man Beast of Darién, Panama, http://www.bigfootencounters.com/creatures/darien.htm. Retrieved Aug. 19, 2014.

16. Almas Sightings, http://unmyst3.blogspot.com/2010/01/almas-sightings.html. Retrieved Aug. 21, 2014.

17. Ibid.

18. Myra Shackley, *Still Living? Yeti, Sasquatch and the Neanderthal Enigma* (New York: Thames and Hudson, 1983), pp. 103-4.

19. Almasti and Kaptars of Russia and Mongolia, http://www.bcscc.ca/almasti.htm. Retrieved Aug. 21, 2014.

20. Shackley, p. 82.

21. Almas (cryptozoology); George M. Eberhart, *Mysterious Creatures: A Guide to Cryptozoology* (Santa Barbara, CA: ABC-CLIO, 2002), p. 266.

22. Shackley, p. 64.

23. Michael Cremo and Richard L. Thompson, *The Hidden History of the Human Race: Major Scientific Coverup Exposed* (Alachua, FL: Bhaktivedanta Book Publishing, 1999), p. 174.

24. Pangboche Hand, http://en.wikipedia.org/wiki/Pangboche_Hand; Jeremy Fugleberg, "The Strange Saga of the Stolen Yeti Hand," http://www.atlasobscura.com/articles/saga-of-the-yeti-hand. Both retrieved Aug. 21, 2014.

25. Pangboche Hand; Fugleberg.

26. Joanna Jolly, " 'Yeti hand' replica to be returned to Nepal monastery," http://www.bbc.co.uk/news/world-south-asia-13228780. Retrieved Aug. 21, 2014.

27. In the mountains of Ingushetia border guards caught a creature similar to "Bigfoot," http://www.interfax.ru/russia/224287; Yeti, http://en.wikipedia.org/wiki/Yeti. Both retrieved Aug. 21, 2014.

Chapter 11

1. Lindsay, "Why No Bigfoot Bones and Bodies?"

2. Bigfoot Hotspot Radio, https://www.facebook.com/BigfootHotspotRadio/posts/381498625286085. Retrieved Aug. 21, 2014.

3. Forest Service Contacts for LWCF, http://www.fs.fed.us. Retrieved Aug. 21, 2014.

4. Bigfoot Hotspot Radio.

5. Ibid.

6. Ibid.

7. Steve Kulls, "Bigfoot Ballyhoo/Linda Newton-Perry," http://squatchdetective.weebly.com/hall-of-shame---bigfoot-ballyhoo--linda-newton-perry.html. Retrieved Aug. 21, 2014.

8. Guy Edwards, "OREGON DMV: We will no longer suspend Drivers Licenses for Reporting Bigfoot/Sasquatch," http://www.bigfootlunchclub.com/search/label/linda%20newton%20perry. Retrieved Aug. 21, 2014.

9. William Jevning, "Cover up in Oregon?" http://jevningresearch.blogspot.com/2011_12_01_archive.html. Retrieved Aug. 21, 2014.

10. Lindsay, "Why No Bigfoot Bones and Bodies?"

11. Lindsay, "Why Has No Hunter Ever Killed a Bigfoot?"

12. Crowe, "Where Are the Bigfoot Bones?"

13. Lindsay, "Why No Bigfoot Bones and Bodies?"

14. Berry, p. 50.

15. Fort Lewis, Washington, http://www.bigfootencounters.com/sbs/ftlewis1.htm. Retrieved August 22, 2014.

16. Ibid.

17. Ibid.

18. Craig Woolheater, "Watch For Sasquatch On Road," http://cryptomundo.com/bigfoot-report/watch-for-sasquatch-on-road. Retrieved Aug. 22, 2014.

19. *Amarillo Globe-News* obituary, Jan. 22, 2008.

20. Washington IBS Sightings Reports.

21. Crowe, "Where Are the Bigfoot Bones?"

22. Lindsay, "Why No Bigfoot Bones and Bodies?"

23. Ibid.

24. Shawn Evidence, "Anonymous Former National Guardsman Claims Mount St. Helens Burnt Bigfoot Story Happened," http://bigfootevidence.blogspot.com/2012/09/anonymous-former-national-guardsman.html. Retrieved Aug. 22, 2014.

25. Ibid.

26. Shawn Evidence, "Airman From 1980s Comes Forward To Talk About Mount St. Helens Burnt Bigfoot Event," http://bigfootevidence.blogspot.com/2012/09/airman-from-1980s-comes-forward-to-talk.html#comment-form. Retrieved Aug. 22, 2014.

27. Scott Brown, "Interview: Cryptozoologist Loren Coleman," trueslant.com/scottbowen/2010/05/27/interview-cryptozoologist-loren-coleman-part-1. Retrieved Aug. 22, 2014.

28. Lindsay, "Why No Bigfoot Bones and Bodies?"

29. Washington IBS Sightings Reports.

30. IBS Report #619, http://www.mid-americabigfoot.com/forums/viewtopic.php?f=167&t=5083&start=440. Retrieved Aug. 22, 2014.

31. USA Place Names.

32. Lindsay, "Why No Bigfoot Bones and Bodies?"

33. GCBRO, http://www.gcbro.com/INmartin0001.html. Retrieved Aug. 22, 2014.

34. Ibid., http://www.gcbro.com/AZmaricopa0001.html. Retrieved Aug. 22, 2014.

35. Lindsay, "Why No Bigfoot Bones and Bodies?"

36. Washington IBS Sightings Reports.

37. Shawn Evidence, "Bigfoot injured by a forest fire was taken away and hidden by the authorities, not even Robert Lindsay can top this story," http://bigfootevidence.blogspot.com/2011/07/bigfoot-injured-by-forest-fire-was.html. Retrieved Aug. 22, 2014.

38. Lindsay, "Why No Bigfoot Bones and Bodies?"

39. Craig Woolheater, "Government Conspiracies and Bigfoot or Do You Smell Something Burning?" http://cryptomundo.com/bigfoot-report/bigfoot-burning. Retrieved Aug. 22, 2014.

40. Bigfoot Shootings.

41. Lindsay, "Why No Bigfoot Bones and Bodies?"

42. Ibid.

43. BFRO Report #7241, http://www.bfro.net/GDB/show_report.asp?id=7241. Retrieved Aug. 22, 2014.

44. Oregonbigfoot.com File #00767, http://www.oregonbigfoot.com/report_detail.php?id=00767. Retrieved Aug. 22, 2014.

45. Search Underway for Loose Primate in Campbell County, WATE-TV Channel 6 (Knoxville, TN), Oct. 20, 2003.

46, Michael Marshall, "Vegetarian orang-utans eat world's cutest animal," http://www.newscientist.com/article/dn21364-vegetarian-orangutans-eat-worlds-cutest-animal.html#.U_h5qWNtHt8. Retrieved Aug. 23, 2014.

47. Is a Skunk Ape Loose in Campbell County? WATE-TV Channel 6, Oct. 22, 2003.

48. "Skunk Ape" DNA Samples To Be Tested, http://www.hotspotsz.com/Skunk_Ape_DNA_Samples_To_Be_Tested_%28Article-10855%29.html. Retrieved Aug. 23, 2014.

49. Loren Coleman, "Has a Skunk Ape Been Killed in TN?" http://www.lorencoleman.com/skunk_ape.html. Retrieved Aug. 23, 2014.

50. Ibid.

51. Lindsay, "Why No Bigfoot Bones and Bodies?"

52. Crowe, "Where Are the Bigfoot Bones?"

53. Lindsay, "Why No Bigfoot Bones and Bodies?"

54. Chemical and Biological Engineering Department, http://www.sdsmt.edu/cbe. Retrieved Aug. 23, 2014.

55. Sharon Hill, "Altoona PA Bigfoot shooting rumor—Case closed," http://doubtfulnews.com/2013/05/altoona-pa-bigfoot-shooting-rumor-case-closed. Retrieved Aug. 24, 2014.

56. James Nye, "Was a Bigfoot shot and killed in rural Pennsylvania?" *Daily Mail* (London), May 29, 2013.

57. Ibid.

58. *The Daily American,* June 1, 2013.

59. Crowe, "Where Are the Bigfoot Bones?"

60. IBS Report #2617, http://www.mid-americabigfoot.com/forums/viewtopic.php?f=167&t=4097. Retrieved Aug. 24, 2014.

61. Crowe, "Where Are the Bigfoot Bones?"

62. Ibid.

63. Ibid.

64. Personal communication with the author, Aug. 24, 2014.

65. James Hagenbruner, "Rumors common, sasquatch aren't." *Billings Gazette,* Oct. 22, 2001.

66. Crowe, "Where Are the Bigfoot Bones?"

67. Michael Krein email, posted to Bigfoot@yahoogroups.com on April 9, 2014.

68. In U.S., 42% Believe Creationist View of Human Origins, http://www.gallup.com/poll/170822/believe-creationist-view-human-origins.aspx. Retrieved Aug. 24, 2014.

69. Republicans, Democrats Differ on Creationism, http://www.gallup.com/poll/108226/Republicans-Democrats-Differ-Creationism.aspx. Retrieved Aug. 24, 2014.

70. Daniel Benoit, "Government & Bigfoot," http://www.ecbro.com/blog-e7325-GOVERNMENT-amp-BIGFOOT.html. Retrieved Aug. 24, 2014.

71. Fred Clark, "Mitt Romney tells 533 lies in 30 weeks, Steve Benen documents them,"

http://www.patheos.com/blogs/slacktivist/2012/08/29/mitt-romney-tells-533-lies-in-30-weeks-steve-benen-documents-them/. Retrieved Aug. 25, 2014.

72. Doug Tarrant's Experiences Shared with Dmitri Bayanov, http://www.bigfootencounters.com/stories/doug.htm. Retrieved Aug. 25, 2014. Used by permission.

73. Ibid.

74. Ibid.

75. Re. Clyde Beatty Circus attempts to import Yeti during Scopes Trial, http://www.cryptozoology.com/forum/topic_view_thread.php?tid=2&pid=526837. Retrieved Aug. 25, 2014.

75. Clyde Beatty Biography, http://www.imdb.com/name/nm0064197/bio. Retrieved Aug. 25, 2014.

76. Email to William Jevning, quoted in "Minnesota Iceman," http://jevningresearch.blogspot.com/2011_12_01_archive.html. Retrieved Aug. 25, 2014.

77. Doug Tarrant, "The Game of futility," ForestGiants@yahoogroups.com. Retrieved July 24, 2014.

78. Ibid.

79. Ibid.

80. Ibid.

81. Doug Tarrant, "Bringing Bigfoot to You," http://www.georgiabigfootsociety.com/contactingBigfoot.html. Retrieved Aug. 25, 2014.

Conclusion

1. Crowe, "Where Are the Bigfoot Bones?"

2. Lindsay, "Why No Bigfoot Bones or Bodies?"

3. Lindsay, "Why Has No Hunter Ever Shot and Killed a Bigfoot?"

4. Lindsay, "Why No Bigfoot Bones or Bodies?"

5. Ibid.

STILL ON THE TRACK OF UNKNOWN ANIMALS

The Centre for Fortean Zoology, or CFZ, is a non profit-making organisation founded in 1992 with the aim of being a clearing house for information, and coordinating research into mystery animals around the world.

We also study out of place animals, rare and aberrant animal behaviour, and Zooform Phenomena; little-understood "things" that appear to be animals, but which are in fact nothing of the sort, and not even alive (at least in the way we understand the term).

Not only are we the biggest organisation of our type in the world, but - or so we like to think - we are the best. We are certainly the only truly global cryptozoological research organisation, and we carry out our investigations using a strictly scientific set of guidelines. We are expanding all the time and looking to recruit new members to help us in our research into mysterious animals and strange creatures across the globe.

Why should you join us? Because, if you are genuinely interested in trying to solve the last great mysteries of Mother Nature, there is nobody better than us with whom to do it.

Members get a four-issue subscription to our journal *Animals & Men.* Each issue contains nearly 100 pages packed with news, articles, letters, research papers, field reports, and even a gossip column! The magazine is Royal Octavo in format with a full colour cover. You also have access to one of the world's largest collections of resource material dealing with cryptozoology and allied disciplines, and people from the CFZ membership regularly take part in fieldwork and expeditions around the world.

The CFZ is managed by a three-man board of trustees, with a non-profit making trust registered with HM Government Stamp Office. The board of trustees is supported by a Permanent Directorate of full and part-time staff, and advised by a Consultancy Board of specialists - many of whom are world-renowned experts in their particular field. We have regional representatives across the UK, the USA, and many other parts of the world, and are affiliated with

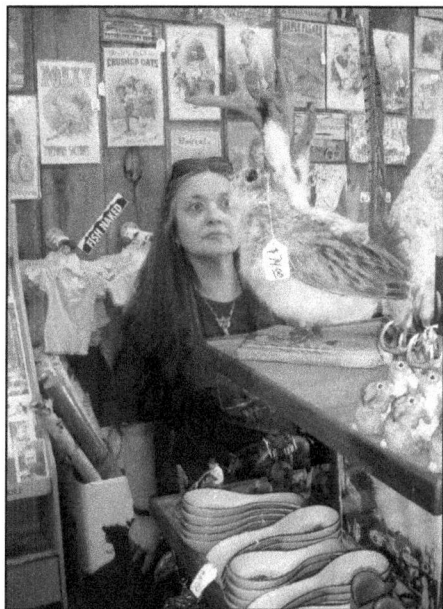

You'll find that the people at the CFZ are friendly and approachable. We have a thriving forum on the website which is the hub of an ever-growing electronic community. You will soon find your feet. Many members of the CFZ Permanent Directorate started off as ordinary members, and now work full-time chasing monsters around the world.

Write to us, e-mail us, or telephone us. The list of future projects on the website is not exhaustive. If you have a good idea for an investigation, please tell us. We may well be able to help.

We are always looking for volunteers to join us. If you see a project that interests you, do not hesitate to get in touch with us. Under certain circumstances we can help provide funding for your trip. If you look on the future projects section of the website, you can see some of the projects that we have pencilled in for the next few years.

In 2003 and 2004 we sent three-man expeditions to Sumatra looking for Orang-Pendek - a semi-legendary bipedal ape. The same three went to Mongolia in 2005. All three members started off merely subscribers to the CFZ magazine. Next time it could be you!

We have no magic sources of income. All our funds come from donations, membership fees, and sales of our publications and merchandise. We are always looking for corporate sponsorship, and other sources of revenue. If you have any ideas for fund-raising please let us know. However, unlike other cryptozoological organisations in the past, we do not live in an intellectual ivory tower. We are not afraid to get our hands dirty, and furthermore we are not one of those organisations where the membership have to raise money so that a privileged few can go on expensive foreign trips. Our research teams, both in the UK and abroad, consist of a mixture of experienced and inexperienced personnel. We are truly a community, and work on the premise that the benefits of CFZ membership are open to all.

Reports of our investigations are published on our website as soon as they are available. Preliminary reports are posted within days of the project finishing.

Each year we publish a 200 page yearbook

We have a thriving YouTube channel, CFZtv, which has well over two hundred self-made documentaries, lecture appearances, and episodes of our monthly webTV show. We have a daily online magazine, which has over a million hits each year.

Each year since 2000 we have held our annual convention - the Weird Weekend. It is three days of lectures, workshops, and excursions. But most importantly it is a chance for members of the CFZ to meet each other, and to talk with the members of the permanent directorate in a relaxed and informal setting and preferably with a pint of beer in one hand. Since 2006 - the Weird Weekend has been bigger and better and held on the third weekend in August in the idyllic rural location of Woolsery in North Devon.

Since relocating to North Devon in 2005 we have become ever more closely involved with other community organisations, and we hope that this trend will continue. We have also worked closely with Police Forces across the UK as consultants for animal mutilation cases, and we intend to forge closer links with the coastguard and other community services. We want to work closely with those who regularly travel into the Bristol Channel, so that if the recent trend of exotic animal visitors to our coastal waters continues, we can be out there as soon as possible.

© Undergroundimages2007

Apart from having been the only Fortean Zoological organisation in the world to have consistently published material on all aspects of the subject for over a decade, we have achieved the following concrete results:

• Disproved the myth relating to the headless so-called sea-serpent carcass of Durgan beach in Cornwall 1975
• Disproved the story

of the 1988 puma skull of Lustleigh Cleave

- Carried out the only in-depth research ever into the mythos of the Cornish Owlman.
- Made the first records of a tropical species of lamprey
- Made the first records of a luminous cave gnat larva in Thailand
- Discovered a possible new species of British mammal - the beech marten
- In 1994-6 carried out the first archival fortean zoological survey of Hong Kong
- In the year 2000, CFZ theories were confirmed when a new species of lizard was added to the British List
- Identified the monster of Martin Mere in Lancashire as a giant wels catfish
- Expanded the known range of Armitage's skink in the Gambia by 80%
- Obtained photographic evidence of the remains of Europe's largest known pike
- Carried out the first ever in-depth study of the ninki-nanka
- Carried out the first attempt to breed Puerto Rican cave snails in captivity
- Were the first European explorers to visit the `lost valley` in Sumatra
- Published the first ever evidence for a new tribe of pygmies in Guyana
- Published the first evidence for a new species of caiman in Guyana

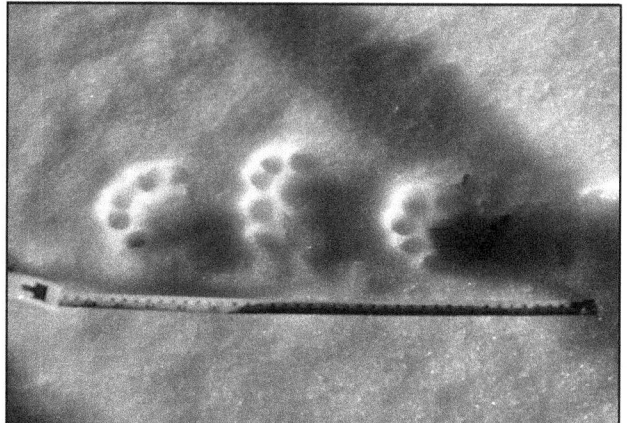

on a monster-haunted lake in Ireland for the first time
- Had a sighting of orang pendek in Sumatra in 2009
- Found leopard hair, subsequently identified by DNA analysis, from rural North Devon in 2010
- Brought back hairs which appear to be from an unknown primate in Sumatra
- Published some of the best evidence ever for the almasty in southern Russia

CFZ Expeditions and Investigations include:

- 1998 Puerto Rico, Florida, Mexico (Chupacabras)
- 1999 Nevada (Bigfoot)
- 2000 Thailand (Naga)
- 2002 Martin Mere (Giant catfish)
- 2002 Cleveland (Wallaby mutilation)
- 2003 Bolam Lake (BHM Reports)

- 2003 Sumatra (Orang Pendek)
- 2003 Texas (Bigfoot; giant snapping turtles)
- 2004 Sumatra (Orang Pendek; cigau, a sabre-toothed cat)
- 2004 Illinois (Black panthers; cicada swarm)
- 2004 Texas (Mystery blue dog)
- Loch Morar (Monster)
- 2004 Puerto Rico (Chupacabras; carnivorous cave snails)
- 2005 Belize (Affiliate expedition for hairy dwarfs)
- 2005 Loch Ness (Monster)
- 2005 Mongolia (Allghoi Khorkhoi aka Mongolian death worm)

- 2006 Gambia (Gambo - Gambian sea monster , Ninki Nanka and Armitage's skink
- 2006 Llangorse Lake (Giant pike, giant eels)
- 2006 Windermere (Giant eels)
- 2007 Coniston Water (Giant eels)
- 2007 Guyana (Giant anaconda, didi, water tiger)
- 2008 Russia (Almasty)
- 2009 Sumatra (Orang pendek)
- 2009 Republic of Ireland (Lake Monster)
- 2010 Texas (Blue Dogs)
- 2010 India (Mande Burung)
- 2011 Sumatra (Orang-pendek)

For details of current membership fees, current expeditions and investigations, and voluntary posts within the CFZ that need your help, please do not hesitate to contact us.

The Centre for Fortean Zoology,
Myrtle Cottage,
Woolfardisworthy,
Bideford, North Devon
EX39 5QR

Telephone 01237 431413
Fax+44 (0)7006-074-925
eMail info@cfz.org.uk

Websites:

www.cfz.org.uk
www.weirdweekend.org

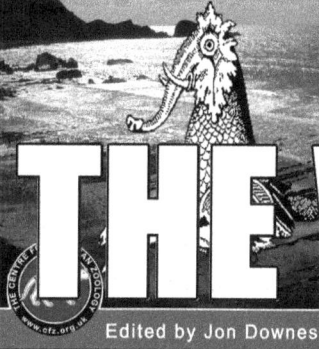

ANIMALS & MEN
ISSUES 16-20
THE JOURNAL OF THE CENTRE FOR FORTEAN ZOOLOGY
NEW HORIZONS
Edited by Jon Downes

BIG CATS
LOOSE IN BRITAIN

PREDATOR DEATHMATCH
NICK MOLLOY
WITH ILLUSTRATIONS BY ANTHONY WALLIS

THE WORLD'S WEIRDEST

TER!
PHENOMENA

Edited by
Jonathan Downes and Richard Freeman

FOREWORD BY Dr. KARL SHUKER

PUBLISHING

A DAINTREE DIARY
Tales from Travels Daintree
tropical North

CARL PORTMAN

STAR STEEDS

THE COLLECTED POEMS
Dr Karl P.N. Shuker

STRANGELYSTRANGE
ly normal

an anthology of writings by
ANDY ROBERTS

COMPANY

HOW TO START A PUBLISHING EMPIRE

Unlike most mainstream publishers, we have a non-commercial remit, and our mission statement claims that "we publish books because they deserve to be published, not because we think that we can make money out of them". Our motto is the Latin Tag *Pro bona causa facimus* (we do it for good reason), a slogan taken from a children's book *The Case of the Silver Egg* by the late Desmond Skirrow.

WIKIPEDIA: "The first book published was in 1988. *Take this Brother may it Serve you Well* was a guide to Beatles bootlegs by Jonathan Downes. It sold quite well, but was hampered by very poor production values, being photocopied, and held together by a plastic clip binder. In 1988 A5 clip binders were hard to get hold of, so the publishers took A4 binders and cut them in half with a hacksaw. It now reaches surprisingly high prices second hand.

The production quality improved slightly over the years, and after 1999 all the books produced were ringbound with laminated colour covers. In 2004, however, they signed an agreement with Lightning Source, and all books are now produced perfect bound, with full colour covers."

Until 2010 all our books, the majority of which are/were on the subject of mystery animals and allied disciplines, were published by `CFZ Press`, the publishing arm of the Centre for Fortean Zoology (CFZ), and we urged our readers and followers to draw a discreet veil over the books that we published that were completely off topic to the CFZ.

However, in 2010 we decided that enough was enough and launched a second imprint, `Fortean Words` which aims to cover a wide range of non animal-related esoteric subjects. Other imprints will be launched as and when we feel like it, however the basic ethos of the company remains the same: Our job is to publish books and magazines that we feel are worth publishing, whether or not they are going to sell. Money is, after all - as my dear old Mama once told me - a rather vulgar subject, and she would be rolling in her grave if she thought that her eldest son was somehow in `trade`.

Luckily, so far our tastes have turned out not to be that rarified after all, and we have sold far more books than anyone ever thought that we would, so there is a moral in there somewhere...

Jon Downes,
Woolsery, North Devon
July 2010

Other Books in Print

by Downes, Jonathan
The Smaller Mystery Carnivores of the Westcountry by Downes, Jonathan
CFZ EXPEDITION REPORT: Gambia 2006 by Richard Freeman *et al*, Shuker, Karl (fwd)
The Owlman and Others by Jonathan Downes
The Blackdown Mystery by Downes, Jonathan
Big Cats in Britain Yearbook 2006 by Fraser, Mark (Ed)
Fragrant Harbours - Distant Rivers by Downes, John T
Only Fools and Goatsuckers by Downes, Jonathan
Monster of the Mere by Jonathan Downes
Dragons:More than a Myth by Freeman, Richard Alan
Granfer's Bible Stories by Downes, John Tweddell
Monster Hunter by Downes, Jonathan

TRADE MARK
BEWARE OF IMITATIONS
®

CFZ CLASSICS

CFZ Classics is a new venture for us. There are many seminal works that are either unavailable today, or not available with the production values which we would like to see. So, following the old adage that if you want to get something done do it yourself, this is exactly what we have done.

Desiderius Erasmus Roterodamus (b. October 18th 1466, d. July 2nd 1536) said: "When I have a little money, I buy books; and if I have any left, I buy food and clothes," and we are much the same. Only, we are in the lucky position of being able to share our books with the wider world. CFZ Classics is a conduit through which we cannot just re-issue titles which we feel still have much to offer the cryptozoological and Fortean research communities of the 21st Century, but we are adding footnotes, supplementary essays, and other material where we deem it appropriate.

Headhunters of The Amazon by Fritz W Up de Graff (1902)

Fortean Words

The Centre for Fortean Zoology has for several years led the field in Fortean publishing. CFZ Press is the only publishing company specialising in books on monsters and mystery animals. CFZ Press has published more books on this subject than any other company in history and has attracted such well known authors as Andy Roberts, Nick Redfern, Michael Newton, Dr Karl Shuker, Neil Arnold, Dr Darren Naish, Jon Downes, Ken Gerhard and Richard Freeman.

Now CFZ Press are launching a new imprint. Fortean Words is a new line of books dealing with Fortean subjects other than cryptozoology, which is - after all - the subject the CFZ are best known for. Fortean Words is being launched with a spectacular multi-volume series called *Haunted Skies* which covers British UFO sightings between 1940 and 2010. Former policeman John Hanson and his long-suffering partner Dawn Holloway have compiled a peerless library of sighting reports, many that have not been made public before.

Other books include a look at the Berwyn Mountains UFO case by renowned Fortean Andy Roberts and a series of forthcoming books by transatlantic researcher Nick Redfern. CFZ Press are dedicated to maintaining the fine quality of their works with Fortean Words. New authors tackling new subjects will always be encouraged, and we hope that our books will continue to be as ground-breaking and popular as ever.

Haunted Skies Volume One 1940-1959 by John Hanson and Dawn Holloway
Haunted Skies Volume Two 1960-1965 by John Hanson and Dawn Holloway
Haunted Skies Volume Three 1965-1967 by John Hanson and Dawn Holloway
Haunted Skies Volume Four 1968-1971 by John Hanson and Dawn Holloway
Haunted Skies Volume Five 1972-1974 by John Hanson and Dawn Holloway
Haunted Skies Volume Six 1975-1977 by John Hanson and Dawn Holloway
Grave Concerns by Kai Roberts

Police and the Paranormal by Andy Owens
Dead of Night by Lee Walker
Space Girl Dead on Spaghetti Junction - an anthology by Nick Redfern
I Fort the Lore - an anthology by Paul Screeton
UFO Down - the Berwyn Mountains UFO Crash by Andy Roberts
The Grail by Ronan Coghlan
UFO Warminster - Cradle of Contract by Kevin Goodman
Quest for the Hexham Heads by Paul Screeton

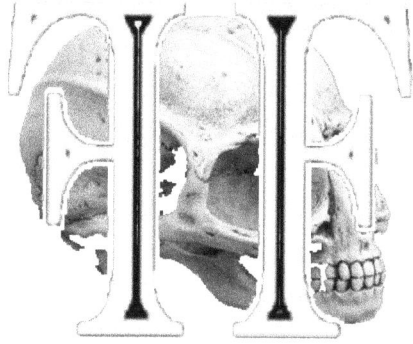

Fortean Fiction

J ust before Christmas 2011, we launched our third imprint, this time dedicated to - let's see if you guessed it from the title - fictional books with a Fortean or cryptozoological theme. We have published a few fictional books in the past, but now think that because of our rising reputation as publishers of quality Forteana, that a dedicated fiction imprint was the order of the day.

We launched with four titles:

Green Unpleasant Land by Richard Freeman
Left Behind by Harriet Wadham
Dark Ness by Tabitca Cope
Snap! By Steven Bredice
Death on Dartmoor by Di Francis
Dark Wear by Tabitca Cope
Hyakymonogatari Book 1 by Richard Freeman

www.ingramcontent.com/pod-product-compliance
Lightning Source LLC
Chambersburg PA
CBHW062224270326
41930CB00009B/1860